Advances in Japanese Business and Economics

Volume 3

Advances in Japanese Business and Economics (AJBE) showcases the work of Japanese and non-Japanese scholars researching the Japanese economy and Japanese businesses. Published in English, the series highlights for a global readership the unique perspectives of Japan's most distinguished and emerging scholars of business and economics. It covers research of either theoretical or empirical nature, in both authored and edited volumes, regardless of the sub-discipline or geographical coverage, including, but not limited to, such topics as macroeconomics, microeconomics, industrial relations, innovation, regional development, entrepreneurship, international trade, globalization, financial markets, technology management, and business strategy. At the same time, as a series of volumes written by Japanese and non-Japanese scholars studying Japan, it includes research on the issues of the Japanese economy, industry, management practice, and policy, such as the economic policies and business innovations before and after the Japanese "bubble" burst in the 1990s.

AJBE endeavors to overcome a historical deficit in the dissemination of Japanese economic theory, research methodology, and analysis. The volumes in the series contribute not only to a deeper understanding of Japanese business and economics but to revealing underlying universal principles.

Overseen by a panel of renowned scholars led by Editor-in-Chief Professor Ryuzo Sato, AJBE employs a single-blind review process in which the Editor-in-Chief, together with the Managing Editors and specialized scholars designated by the Editor-in-Chief or Managing Editors, rigorously reviews each proposal and manuscript to ensure that every submission is a valuable contribution to the global scholarly readership.

All books and chapters in AJBE are indexed in Scopus.

Hiroshi Yoshikawa

Reconstruction of Macroeconomics: Methods of Statistical Physics, and Keynes' Principle of Effective Demand

Springer

Hiroshi Yoshikawa
University of Tokyo
Tokyo, Japan

ISSN 2197-8859 ISSN 2197-8867 (electronic)
Advances in Japanese Business and Economics
ISBN 978-981-19-5263-0 ISBN 978-981-19-5264-7 (eBook)
https://doi.org/10.1007/978-981-19-5264-7

Translation from the Japanese language edition: "MAKURO KEIZAIGAKU NO SAIKOCHIKU: KEINZU TO SHUNPETA" by Hiroshi Yoshikawa, © Hiroshi Yoshikawa 2020. Published by Iwanami Shoten, Publishers, Tokyo. All Rights Reserved.

This Springer imprint is published by the registered company Springer Nature Singapore Pte Ltd.
The registered company address is: 152 Beach Road, #21-01/04 Gateway East, Singapore 189721, Singapore

Preface

COVID-19 has been with us for more than two years and has clouded every inch of our economy. In the United States, the unemployment rate which was 4.4% in March 2020 skyrocketed to 14.7% the following month. The growth rate of the world economy abruptly fell to –9.1% (2020Q2). It was the worst drop in growth the world had experienced after the war.

As in the Great Recession following the 2007–2009 global financial crisis, governments around the world expanded expenditures in unison and prevented the downfall of the economy. It was fine. The problem is the wide gulf between such policy actions and mainstream macroeconomics. Fiscal expansion in a time of recession is the ABC's of Keynesian economics, but neoclassical macroeconomics, such as real business cycle theory, does not endorse it.

Modern micro-founded macroeconomics emphasizing the role of expectations allegedly provides monetary policy with solid theoretical foundations for avoiding deflation and achieving healthy growth and a 2% inflation target. The achievement was once hailed as Great Moderation. However, the 2007–2009 global financial crisis turned this confidence to a great disappointment. Facing a zero interest rate bound, the macroeconomics profession has advocated unconventional monetary policies that were supposed to lead the macroeconomy to a stable growth path with the targeted 2% inflation. And yet, in reality, monetary policy failed to achieve this goal. As of March 2022, the U.S. Consumer Price Index rose by 8.5%, a 40-year historical high.

Today's neoclassical macroeconomics, which is taken by the profession to have solid micro-foundations, fails. The reason is simple: The methodology is wrong. The macroeconomy, consisting of many households and firms, cannot be understood by analyzing optimization of micro-agents. In analyzing macro, we must use methods of statistical physics. The spirit is that we must purposely disregard details of micro behavior. Here lies a fundamental difference between macroeconomics and microeconomics. The confusion lasted more than 50 years. It should be the agenda of epistemological study.

Chapter 3 explains that methods of statistical physics provide Keynes's principle of effective demand with robust micro-foundations. It is a great encouragement for me that Professor Robert M. Solow endorsed this analysis. Aggregate demand changes much more swiftly than technology and becomes the effective constraint on the macroeconomy. The profession has purged demand from macroeconomics, and fallen into a state of confusion.

I owe this work to a number of people. I long ago learned Keynesian economics from my mentor, the late James Tobin. His wisdom should still be a guide today. The late Masanao Aoki, a pioneer in this field, introduced me to methods of statistical physics. Professor Katsuhito Iwai has always inspired me with his disequilibrium dynamics.

I have worked with a group of econophysicists, professors Hideaki Aoyama, Hiroshi Iyetomi, Yoshihisa Fujiwara, Yuichi Ikeda, and Wataru Souma, and learned a great deal from them. Some of the materials presented in this book are outputs of my joint works with them. Professor Didier Sornette, one of the world leaders in econophysics, kindly read a part of the manuscript and gave me useful comments which saved me from errors. Professor Ryuzo Sato has always encouraged me. This time, I learned from his early works on technical progress. Dr. Yoshiyuki Arata performed an illustrative simulation shown in Chap. 3.

Ms. Shuko Miyakawa so ably prepared the manuscript and figures. Mr. Hiroshi Takahashi and Ms. Juno Kawakami as editors professionally transformed the manuscript to book form.

I dedicate this book to my wife Setsuko, who has supported me for nearly a half-century.

Tokyo, Japan Hiroshi Yoshikawa
April 2022

Acknowledgments

I am grateful to Professor Daniel Kahneman and the American Economic Association for allowing me to reprint some paragraphs from D. Kahneman, J. Knetsch, and R. Thaler (1986) in this book. My thanks also go to the Econometric Society for allowing me to use Fig. 3, "UK business cycle index (1855–1877)" from Eugen Slutzky (1937); the American Economic Association for Fig. 1 from Nicholas Gregory Mankiw (1989); J. C. Fisher and R. H. Pry for Fig. 9 (b) (1971); the University of Texas Press for Chart III-2 from W. W. Rostow (1978); the Cambridge University Press for Fig. 4.3 from R. Mantegna and H. E. Stanley (2000); and Taylor & Francis for Table 1 from M. Newman (2005).

I have also obtained full reprint permission for all citations and reprints of illustrations from works either single-authored or co-authored by myself.

I am grateful to the Research Institute of Economy, Trade, and Industry (RIETI), Tokyo, for support in preparing this book. This work is partially supported by Grants-in-Aid for Scientific Research (KAKENHI), Grant Numbers 20H02391 and 20H01480, by the Japan Society for the Promotion of Science (JSPS).

Praise for *Reconstruction of Macroeconomics: Methods of Statistical Physics, and Keynes' Principle of Effective Demand*

"It (Chap. 3) captures analytically a good part of the intuition that underlies the Keynesian economics of people like Tobin and me."

—Robert Solow, *Emeritus Institute Professor of Economics, Massachusetts Institute of Technology, Nobel Laureate in Economics, 1987*

"Professor Hiroshi Yoshikawa provides a unique synthesis of statistical physics and macro-economic theory in order to confront the dismal failure in economics and in finance to understand how an economy or a financial market works, given the heterogeneous decision making of many different individual interacting actors. Economics has failed in this regard with the naive and often misleading concept of "representative agents". The author presents many insights on the historical development, concepts, and errors made by the most illustrious economists in the past. This book should be essential readings for any economics students as well as academic researchers and policy makers, who should learn to bring back good-sense thinking in their impactful decisions."

—Didier Sornette, *Professor on the Chair of Entrepreneurial Risks at the Swiss Federal Institute of Technology Zurich (ETH Zurich)*

Introduction

> *What impedes the progress of science is not the ignorance of the layman, but the lack of awareness of the mission and essence of science by the scientist himself!*
>
> *(Torahiko Terada, 1933).*

Over the last 50 years, macroeconomics has drastically changed. Born in the midst of the Great Depression, it was once synonymous with Keynesian economics. However, under the flag of micro-foundations, it has turned to neoclassical macroeconomics.

Modern micro-founded macroeconomics is proud of its solid micro-foundations by which the profession understands that optimization of a micro-economic agent such as the household or firm is explicitly analyzed. In this book, we maintain that standard micro-foundations are on the wrong track. Ironically, modern macroeconomics, which emphasizes the significance of micro-optimization so much, takes it too carelessly.

The fundamental problem is that modern micro-founded macroeconomics rests on representative agent assumptions. In the case of real business cycle (RBC) theory and endogenous growth theory, this is obvious, for a literally single representative household/firm is assumed. In contrast, in the case of Lucas's rational expectations model, labor search theory, and more recent mean field games, the problem is much subtler because in these models, apparently heterogeneous agents are assumed. However, economic agents differ only to the extent that the realization of relevant stochastic variables is different. They are assumed to face a common stochastic distribution with which they optimize. We contend that the assumption that all the economic agents share a common stochastic environment, namely, the assumption of representative constraints, is wrong. It is essentially nothing but a stochastic version of representative agent assumptions, not worthy of micro-foundations.

In modern micro-founded macroeconomics, aggregate demand has been purged as a determinant of the level of macroeconomic activity. This makes macroeconomics more or less Walrasian if on the surface it is different from Walras's original framework. The tragedy is that the macroeconomy in which we happen to live is not Walrasian. In contrast to financial markets, prices play only a limited role in the real economy.

We provide Keynes's principle of effective demand with a solid micro-foundation based on methods of statistical physics. Aggregate demand is the fundamental factor that governs the matching of workers and firms in labor markets.

Demand plays a central role not only in the short run but also in the process of economic growth because in advanced economies, saturation of demand for the existing goods and services constrains economic growth. Growth is generated by product innovations. This is where Keynes and Schumpeter meet.

The role of the financial market must be reconsidered. No doubt a well-developed financial market contributes to economic development through efficient resource allocation. However, it occasionally disturbs the macroeconomy. Efficient market hypotheses must be discarded as universal theory; "bubble" is a meaningful concept. We argue that the notion of "efficiency" differs fundamentally in the real economy and the financial market.

Macroeconomics requires different methods than microeconomics, that is, methods of statistical physics. Macroeconomics must change.

Contents

About the Author

Hiroshi Yoshikawa is professor emeritus at the University of Tokyo and honorary president of the Policy Research Institute, Ministry of Finance, Japan. His area of expertise is macroeconomics and the Japanese economy. He obtained a B.A. in economics from the University of Tokyo in 1974 and a Ph.D. in economics at Yale University in 1978. His Ph.D. advisor was the late James Tobin. Professor Yoshikawa's key publications include *Macroeconomics and the Japanese Economy* (Oxford University Press 1995), *Reconstructing Macroeconomics: A Perspective from Statistical Physics and Combinatorial Stochastic Processes* (Cambridge University Press 2007), with Masanao Aoki; *Macro-Econophysics* (Cambridge University Press 2017), with Hideaki Aoyama, Yoshi Fujiwara, Yuichi Ikeda, Hiroshi Iyetomi, and Wataru Souma; and *Complexity, Heterogeneity, and the Methods of Statistical Physics in Economics* (Springer 2020), with Hideaki Aoyama and Yuji Aruka. Professor Yoshikawa was awarded the Medal with Purple Ribbon from the Japanese government in 2010.

Chapter 1
Why Are the Standard Microfoundations Wrong?

1.1 Beginnings

The term "macroeconomics," the subject of this book, came into use after World War II.[1] The "income analysis" instead of macroeconomics was used in Samuelson's *Economics* first published in 1948, which was a worldwide bestseller long after the war. Whatever the term, macroeconomics once aimed to explain how the level of aggregate income or gross domestic product (GDP) is determined based on the Keynesian principle of effective demand and what fiscal and monetary policies should be taken to achieve full employment. Macroeconomics was synonymous with Keynesian economics, apart from 'growth theory', which analyzes long-term economic growth.

For two decades after the war until the end of the 1960s, most economists accepted the idea that economics was dual on two pillars, namely, microeconomics, also known as "price theory", and macroeconomics (Keynesian economics). Samuelson's so-called *neoclassical synthesis* treated the relationship between the two as a dichotomy between neoclassical theory, which was meant to analyze resource allocation under the state of "full employment", and Keynesian economics, which analyzed the economy with unemployment. Our economies occasionally suffer from severe recessions and high unemployment, and in such cases, as Keynesian economics teaches, it is necessary for the government to achieve full employment as quickly as possible using appropriate fiscal/monetary policy. Once full employment is achieved, the efficient allocation of resources is also important, and this is where the neoclassical equilibrium theory can be used. This was the idea of the "neoclassical synthesis" and

---

[1] The origin of the term "macroeconomics" is not altogether clear. Sandomo (2011, 339) makes the following remark: "The origin of the terms *micro-* and *macroeconomics* is not quite clear, but much suggests that they are due to the Norwegian economist Ragnar Frisch who in a 1933 article used the concepts of micro- and macrodynamics. The words *microeconomics* and *macroeconomics* were later used by the research group surrounding the Dutch economist Jan Tinbergen, who in his early work was strongly influenced by Frisch".

H. Yoshikawa, *Reconstruction of Macroeconomics: Methods of Statistical Physics, and Keynes' Principle of Effective Demand*, Advances in Japanese Business and Economics 3, https://doi.org/10.1007/978-981-19-5264-7_1

was widely accepted by the academic community. All over the world, economics was taught, and textbooks were written on the basis of these ideas.

However, in the mid-1960s, the neoclassical synthesis of micro/macroeconomics was gradually eroded. In his review of Samuelson's collected papers, Arrow criticized neoclassical synthesis as a mere common sense devoid of rigor (Arrow, 1967).

> Let $W(g)$ be the Walrasian system and $K(g)$ be the "true" system, where g is a parameter representing government action (fiscal or monetary). The Samuelson-Keynes view of the world is that full employment is a valid proposition in $K(g)$ only for special value of g, whereas full employment holds in $W(g)$ for all g. If g^* is such that full employment holds in $K(g^*)$, can it be true that theorems valid in $W(g^*)$ are also valid in $K(g^*)$? Obviously, it is not true that the two systems respond similarly to changes in g, since full employment remains valid in one but not in the other. However, one might argue that the appropriate criterion is that the response of the economic behavior predicted by $K(g^*)$ to shifts in parameters of the system other than government actions (for example, technical change, shifts in demand for exports) be quantitatively, or at least qualitatively, the same as for $W(g^*)$. There is no a priori reason to believe this, and any brief consideration of possible K-models makes the proposition seem unlikely, except possibly at a more sophisticated level of analysis than has yet been achieved.
>
> For example, if the underlying element in a K-model is a price rigidity of some sort, appropriate government spending or taxation may indeed create full employment; but the resulting equilibrium will not shift in the neoclassical manner to technological change. Only if it is assumed that the price rigidity will somehow soften in the beneficent sun of prolonged full employment can one derive neoclassical results in the long run (Arrow, 1967, 735).

How are we to understand the relationship between Keynesian economics and neoclassical theory? This question, of course, did not begin with Arrow's discussion. It was actively debated immediately after the publication of *General Theory* in February 1936. *The Quarterly Journal of Economics*, for example, published a special issue of *General Theory* as early as November 1936, in which J. Viner, D.H. Robertson, F. Taussig and W. Leontief wrote their critical articles. In response to this, in the February 1937 issue of the same journal, Keynes (1937) himself gave a summary of *General Theory* under the title of "The General Theory of Employment". The contribution of Hicks (1936, 37) is also representative of the works immediately following the publication of *General Theory*. Out of this arose the IS/LM model, which is still a staple of macroeconomics at the undergraduate level today. Keynes's own reactions, including his interactions with Hicks and others, are summarized in Volume 14 of the *Collected Works of JMK*.

Subsequently, Modigliani (1944) proposed a model that organizes neoclassical theory and Keynesian economics as economics that analyzes economies with flexible and rigid nominal wages/prices, respectively. That is, if prices/wages change (even downward) flexibly, the neoclassical "full employment" equilibrium is achieved, but for whatever reason, if prices/wages are inflexible (rigid), Keynesian underemployment ensues. This frame of thought is still widely accepted today. The research conducted in the 1980s under the heading of *New Keynesian Economics* (Mankiw & Romer, 1991) was basically concerned with explaining wage/price rigidity in a "rational" way. The dynamic stochastic general equilibrium (DSGE) model understood as the "Keynesian" model today is also called "Keynesian" to the extent that it assumes inflexible wages/prices.

In the 1960s, when Keynesian economics was still widely accepted as macroeconomics, a new research program was born. Namely, Arrow cited earlier and others gave rise to a new program of research called "microeconomic foundations" that is meant to clarify what assumptions Keynesian economics precisely makes about the behavior of microeconomic agents such as households and firms.

In the Keynesian consumption function, consumption is a function of household income. While this may seem plausible to common sense, "income" is actually the product of wages and labor supply, i.e., the product of "price" and "quantity," rather than "price," such as real wages and interest rates. According to Walrasian general equilibrium theory, households facing wages/prices as given simultaneously determine consumption and labor supply. Clower (1965) re-examined the microeconomic foundations of the consumption function and argued that households actually take not only wages but also labor supply as constraints. The notion of *quantity constraint,* which is absent in Walrasian theory, was born. Benassy (1975), Malinvaud (1977), and Negishi (1979) analyzed quantity constraint models to provide Keynesian economics with microfoundations.[2] Negishi's work will be examined in detail in Chap. 3.

All of the microfoundations of macroeconomics mentioned above were concerned with Keynesian economics, but the research program of "microfoundations" soon underwent a major change in direction, bringing a storm of "Keynesian counterrevolution" to macroeconomics.

The starting point was Friedman's (1968) presidential address to the American Economic Association, in which he criticized the then widely accepted Phillips curve. If there is a negative correlation between price/wage changes and the unemployment rate in the "short run", in the "long run" when "expectations" are realized, such correlation ceases to exist, i.e., there is no Phillips curve.[3] Friedman's argument is based on the classical/neoclassical theory of neutral money, which claims that nominal and real variables are independent; the downward sloping Phillips curve contradicts this fundamental theory.

Soon after, the collection of papers compiled by Phelps (1970) had a decisive influence on macroeconomics through the microfoundations of the Phillips curve. Then, the "rational expectations revolution" of Lucas (1972, 73) and the real business cycle theory (RBC) of Kydland and Prescott (1982) literally transformed macroeconomics. It suffices to cite Lucas' (1987) Declaration of Victory as representative of the trends in the academic world.

> The most interesting recent developments in macroeconomic theory seem to me describable as the reincorporation of aggregative problems such as inflation and the business cycle within

---

[2] Meanwhile, Leiyonhufund (1968) interpreted the *General Theory* as unfinished dynamics.

[3] In 2018, the *Journal of Economic Perspectives,* a journal of the American Economic Association, devoted a special issue to commemorating 50 years of Friedman's presidential address (Hall & Sargent, 2018).

> the general framework of 'microeconomic' theory. If these developments succeed, the term 'macroeconomic' will simply disappear from use and the modifier 'micro' will become superfluous. We will simply speak, as did Smith, Ricardo, Marshall and Walras, of economic theory. If we are honest, we will have to face the fact that at any given time there will be phenomena that are well-understood from the point of view of the economic theory we have, and other phenomena that are not. We will be tempted, I am sure, to relieve the discomfort induced by discrepancies between theory and facts by saying that the ill-understood facts are the province of some other, different kind of economic theory. Keynesian 'macroeconomics' was, I think, a surrender (under great duress) to this temptation. (Lucas, 1987, 107–108).

As Lucas suggests, since the 1980s, macroeconomics has been transformed into microfounded macroeconomics with microoptimization at its core.

1.2 Representative Consumer/Firm

Ramsey's (1928) model, once called the "optimal growth model" in the 1960s and understood as a *normative* model to guide the government in their long-term economic planning, is now regarded as a model that describes the behavior of the real economy. The optimal growth path of the Ramsey model is consistent with a decentralized market equilibrium in the absence of market failures (a macro version of the fundamental theorem of welfare economics); the RBC, beginning with Kydland and Prescott (1982), is representative of such an approach. In Blanchard and Fischer (1989), a best-selling textbook on macroeconomics at the graduate level, the Ramsey model is also described in detail as the basic framework of macroeconomics. We discuss it in Chap. 2.

In fact, from the 1980s to the present day, macroeconomic papers —— surprisingly enough, even those on economic development that focus on developing countries all start with maximizing the utility of the representative Ramsey consumer.

Recent developments of macroeconomics actually emphasize the "heterogeneity" in the economy, though. Therefore, some economists may argue that macroeconomics today is not based on the assumption of representative economic agents. Indeed, formally, in both Lucas's rational expectations model and the search model of labor that we will consider in Chap. 3, there are apparently "heterogeneous" micro agents. For that matter, Walras' theory of general equilibrium assumes that consumers and firms are heterogeneous. The heterogeneity of households and firms has also been emphasized in empirical studies of consumption and investment, partly due to the increasing use of micro data (Attanasio, 1999; Caballero, 1999).

However, as we will explain shortly, in these models, heterogeneity is there only as "convenient variation" around the mean that can be analytically handled in model building. There is a tacit assumption that macroeconomic phenomena can be captured by the behavior of representative economic agents. When we consider optimization explicitly, we must necessarily assume representative consumers and representative firms. It is just impossible to analyze the economy by considering the utility maximizations of one hundred million different households.

The approach based on representative consumers has had a significant impact on empirical analysis as well as theory. The analysis based on the Euler equation derived from the utility maximization of representative consumers was initiated by Hall (1978), and it showed an extraordinary "paper multiplier". The RBC, a representative of modern macroeconomics, understands the business cycle as a result of utility maximization of the representative consumer.

Lucas (1987) calculated the "cost" of the business cycle based on a representative consumer's expected utility maximization framework. Even if all the fluctuations in consumption that had occurred in the U.S. economy since World War II were removed, the per capita gain would be merely \$8.50 per American. Thus, economic fluctuations or the business cycle we have experienced since World War II is a minor problem. This was Lucas's conclusion.

As Fair (1989) pointed out, the gains from economic stabilization are very different in a Keynesian world. For example, if the U.S. recession of 1980–82 (the 1982 recession was the worst postwar recession for the U.S. economy until the Great Recession that occurred with the financial crisis following the collapse of Lehman Brothers in the fall of 2008) had been completely eliminated, and the economy had grown at an annualized rate of 3 percent from 1979 to 1982 instead of actual growth rates of −0.2% in 1980, 1.9% in 1981, and −2.5% in 1982, then the gain per American would have been \$2,400, a far cry from the \$8.50 calculated by Lucas.

In addition, in reality, the costs of the business cycle are not borne equally by every person. As is clear when considering unemployment, the costs of recessions are usually greater for weaker members of society, and thus, they are "regressive". This is why recessions are a major social problem in all countries at all times, and they always strongly influence politics.

Lucas also gave a presidential address at the annual meeting of the American Economic Association to the effect that all the problems of the business cycle had been solved (Lucas, 2003). When Lucas gave this address, the problems of the U.S. economy, including subprime mortgages, worsened in the run-up to the financial crisis of 2008–09, which would later be called the Great Recession. The work of Lucas, a leading academic, is emblematic of how far removed from reality current macroeconomics is.

Microfounded macroeconomics, starting from the maximization of the utility of representative consumers, has been embraced by most economists for the past 40 years. However, these trends among economists are actually quite strange if we extend our horizons beyond economics to the broader academic world. As we will explain in Chap. 2, in the natural sciences, including physics, chemistry, and biology, there is a distinction between the analysis of the movement of a single "particle" or "object" and the analysis of the system consisting of a large number of micro units. It is common sense to use a completely different theoretical framework when analyzing macro systems consisting of a large number of micro units. It is certainly true that the behavior of economic agents is based on rational optimization. However, as we will explain later, it is only a rudimentary misunderstanding to think that economic behavior is inherently different from the motion of "inorganic" particles, which are the objects of natural science.

As Lucas's words symbolize, economists have been passionate about integrating microeconomics (the analysis of micro behavior) and macroeconomics (the analysis of macroeconomic phenomena). That is why microfounded macroeconomics has been embraced by many economists. However, the attempt to "incorporate macroeconomics into the general framework of microeconomic theory" (Lucas, 1987, cited above) by way of representative consumer utility maximization is an incorrect research program. The resultant microfounded macroeconomics is nothing more than a castle in the sand.

However, the academic community of orthodox economics, particularly in the United States, has been very firm and is not ready to listen to such criticisms. Of course, criticism of microfounded macroeconomics has not been entirely absent in orthodox economics. Kirman (1992), for example, was scathing in his critique of the core concept of microfounded macroeconomics, "representative economic agents". Hahn and Solow (1995) also criticized microfounded macroeconomics as represented by the RBC based on a model of overlapping generations (OLG). Furthermore, Solow (2004), in a paper entitled "The Tobin Approach to Monetary Economics", warns against "too much microfoundation". However, microfounded macroeconomics is still mainstream, and almost all the papers, albeit couched in terms of the "heterogeneity" of economic agents, essentially start with the utility maximization of representative individuals.

The financial crisis and the global recession that escalated following the collapse of Lehman Brothers on September 15, 2008, threw cold water on the current state of macroeconomics. *The Economist*, July 18–24, 2009, for example, published a special feature entitled "Modern Economic Theory: Where it went wrong, and how the crisis is changing it". Additionally, in June 2009, in the Lionel Robbins Lectures at the London School of Economics, Paul Krugman stated, "Most macroeconomics of the past 30 years has been spectacularly useless at best, and positively harmful at worst." The previously mentioned Lucas (1987) assertion that even if the entire postwar U.S. business cycle were to be removed, the gain per representative consumer would be mere $8.50 so that recessions do not matter is certainly "spectacularly useless at best, and positively harmful at worst". Fortunately, there were no politicians or policymakers who took this nonsense seriously.

In this vein, Stiglitz (2010) also argues in his article titled "Needed: a new economic paradigm" that

> The blame game continues over who is responsible for the worst recession since the Great Depression – the financiers who did such a bad job of managing risk or the regulators who failed to stop them. But the economics profession bears more than a little culpability. ......
>
> It is hard for non-economists to understand how peculiar the predominant macroeconomic models were. Many assumed demand had to equal supply – and that meant there could be no unemployment. (Right now a lot of people are just enjoying an extra dose of leisure; why they are unhappy is a matter for psychiatry, not economics.) Many used "representative agent models" – all individuals were assumed to be identical, and this meant there could be no meaningful financial markets (who would be lending money to whom?). ......
>
> ......

> Changing paradigms is not easy. Too many have invested too much in the wrong models. Lie the Ptolemaic attempts to preserve earth-centric views of the universe, there will be heroic efforts to add complexities and refinements to the standard paradigm. The resulting models will be an improvement and policies based on them may do better, but they too are likely to fail. Nothing less than a paradigm shift will do (Stiglitz, 2010).

Despite this growing critique of microfounded macroeconomics from within orthodox economics, the main reason why it has failed to move the economics majority to use Stiglitz's (2010) words must be because we cannot see the new "paradigm". The main problem is that economics does not know the methodology of "microfoundations" other than the optimization of microeconomic agents. However, there is, in fact, a clear method other than "microoptimization".

What gives macroeconomics its proper "microfoundations" is, in a word, the methodology of *statistical physics*. In economics, there must be a dichotomy between microeconomics, which analyzes microeconomic problems, and macroeconomics, which analyzes macroeconomic phenomena. When the problem is micro, the incentives of individual economic agents matter. Here, optimization is important and meaningful. However, in the case of macroeconomics, the behavior of microactors, especially their interactions, is so complex that it is literally unobservable to third parties and requires a completely different approach. That is the methods of statistical physics. Perhaps surprisingly, it provides Keynes's principle of effective demand, the pillar of macroeconomics, with a solid macro foundation (Chap. 3).

Standard "microfoundations" that analyze in detail the optimizing behavior of microeconomic agents such as consumers and firms necessarily assume "representative economic agents". Most of the real business cycle (RBC) and endogenous growth models literally assume a representative consumer. However, it is just impossible to describe macroeconomic behavior in terms of the behavior of a single "representative" consumer because real consumers and businesses are plainly different in all respects, including preferences, technology and size. Everyone would think so.

However, there is a defense of RBCs by Kiyotaki (2011):

> Note that the representative household is not an assumption; it arises as an implication of constant Negishi weights under complete markets as in Negishi (1960). The aggregate production function is also constructed under the assumption that production is efficient in competitive markets without friction. Therefore, the real business cycle theory does not blindly abstract from firm and household heterogeneity. By assuming that markets are functioning "well," we reduce an otherwise general model to one of the representative agent with an aggregate production function and analyze the business cycle phenomenon in this simplified economy (Kiyotaki, 2011, 197).

Is this a good defense? He is formally correct, but what exactly is the point of such an argument? As we will explain later, there is actually a fundamental problem with Walras' general equilibrium theory in the first place. In his own influential paper, Kiyotaki and Moor (1997), economic agents are subject to borrowing constraints, but in such a world, the above argument does not hold.

The method of statistical physics explained in Chaps. 2 and 3 is necessary because the movement of many microeconomic agents is inevitably accompanied by randomness, which can never be known by a third party analyzing the macroeconomy. Micro randomness cannot be *a priori* expected to cancel each other out. Rather than justifying a model that assumes representative economic agents, Kiyotaki's statement actually highlights serious problems of Walras' general equilibrium theory.

In Walras' theory of general equilibrium, which we will discuss in more detail later, the preferences of individual consumers and the technology of individual firms are all different. Thus, micro agents are heterogeneous. However, under certain assumptions, as Kiyotaki (2011) puts it, it can be analyzed as if there were "representative consumers". This is because Walras' general equilibrium theory makes the unrealistic assumption that the constraints under which each economic agent optimizes are only prices, and therefore, as a result of optimization, the marginal rates of substitution of all economic agents become equal to a common price vector.

As stated in the Introduction, many of the mainstream macroeconomic models, such as Lucas's rational expectations model and the search theory of labor discussed in Chap. 3, assume superficially heterogeneous economic agents. They not only assume but also emphasize the "heterogeneity" of micro agents. Therefore, many economists believe that macroeconomics with standard microfoundations does not assume representative economic agents and that the criticism that the assumption of representative economic agents is unrealistic is beside the point. However, as explained in more detail below, standard macroeconomic models do not correctly model the *real heterogeneity* of microeconomic agents because they differ only to the extent that realizations of relevant stochastic variables are different; they are assumed to face a common stochastic distribution under which they optimize (Krusell & Smith, 1998). These models are a *stochastic version of the representative agent assumption.* In other words, they only assume apparent "heterogeneity" that is convenient for model building. The best way to understand this point is to consider it by way of concrete examples.

1.3 Lucas Model—An Example

In standard microfounded macroeconomics, models such as Lucas's rational expectations model, the search theory of the labor market and the more recent mean field game assume "heterogeneous" microeconomic agents on the surface. However, all economic agents in these models share the same probability distribution and differ from each other only in the sense that the *realizations* of the random variables differ. The same probability distribution, i.e., a "representative probability distribution" common to all economic agents, is assumed.

Macro distribution, of course, exists for any economic variable. For example, there is a distribution of wages in the economy. However, such a macro distribution is important to the Ministry of Labor, say. It is not relevant for individual workers looking for jobs. There is no such thing as a common constraint for them because the constraints of individual economic agents are all heterogeneous, including the probability distribution facing them.

Phelps (1970) introduced the so-called islands model in which micro agents share the same probability distribution as a constraint for their optimization. In Lucas's (1972, 1973) rational expectations model, endogenous growth theory (Aghion & Howitt, 1992), and search theory (Lucas & Prescott, 1994; Mortensen & Pissarides, 1994), the same assumptions are made, albeit in different forms.

As an example of Phelps' islands model, let us specifically consider Lucas' rational expectations model, which transformed macroeconomics in the 1970s (Lucas, 1973). The core of Lucas' model is the "aggregate supply function." It is derived from the micro supply function of the individual market as follows.

A number of different firms exist in a separate market indexed by z. The cyclical output $y_t(z)$ of market z, which varies with the business cycle, depends on the perceived relative price of the firms in this market—the only signal the firms receive as a constraint is price!

Now, even if *ex post* it is wrong, if a firm believes *ex ante* that the relative price of the good it produces has increased, it will increase its output, and *vice versa.* Equation (3) in Lucas's (1973) paper expresses the output $y_t(z)$ of the market z, *which* is thus determined.[4]

$$y_t(z) = \gamma[P_t(z) - E(P_t|I_t(z))] \qquad (\gamma > 0) \tag{1.1}$$

where $P_t(z)$ is the actual price in market z in period t (logarithmic value), while $E(P_t \mid I_t(z))$ is the conditional expected value for the macro average price P_t formed by the firm in market z based on the local information $I_t(z)$ available in period t. If we note that P_t is logarithmic, we can see that the parentheses represent the "perceived relative price" of the good in market z.

The prior distribution that firms in each market have for the macro price P_t is common to all markets z and is a normal distribution with mean P_t and varianceσ^2. It is easy to accept the assumption that all firms share a common distribution, but as we will explain later, this is where the biggest problem lies. The reason why economists accept this assumption is that they lose sight of the position of individual microeconomic agents and lapse into a bird's eye view of the macroeconomy from the sky.

Now, assume that the price $P_t(z)$ of an individual good supplied in market z deviates from the macro average price P_t by z.

$$P_t(z) = P_t + z \tag{1.2}$$

In other words, the micromarket index z *is* equated with the deviation from the average price (the deviation expressed as a percentage, since P is logarithmic) z.[5]

---

[4] We have simplified Eq. (3) in Lucas (1973) to explain only the essentials.

[5] Thus, in this model, z is not a firm/industry-specific index, for example, the steel industry. The market-firm is indexed according to the size of the deviation from the average price in each period (the change in relative prices). In general, the firms that belong to a given z are different each period.

For a "microshock" z, it is assumed that it follows a normal distribution with mean zero and variance τ^2 that is independent of the distribution of P. Note that all industries and firms have the same variance in terms of relative prices. Individual firms are different in the sense that the realizations of z are different, but they face exactly the same distribution. It is, therefore, effectively a representative firm in that it is constrained by a single distribution. This is a routine assumption in standard microfounded microeconomics.

In any case, under this assumption, the information $I_t(z)$ needed to calculate $E(P_t \mid I_t(z))$ is $P_t(z)$ observed in the individual market z and the prior $\overline{P_t}$ for the average price P_t. As a result,

$$E(P_t|I_t(z)) = E(P_t|P_t(z), \overline{P_t}) = (1-\theta)P_t(z) + \theta\overline{P_t} \tag{1.3}$$

where the weight θ for $\overline{P_t}$ is

$$\theta = \frac{\tau^2}{\sigma^2 + \tau^2} \tag{1.4}$$

That is, the conditional expected value or "rational expectations", $E(P_t \mid I_t(z))$ is a weighted average of the local price $P_t(z)$ observed in the individual market z and the prior average price $\overline{P_t}$. The weight depends on the variances of the "microshock" z, τ^2 and the macro average price P_t, σ^2. The larger the variance is, the smaller the information value of the variable, and as a consequence, the smaller its weight.

From (1.1) and (1.3), the micro supply function y of the individual market z is as follows:

$$y_t(z) = \theta\gamma[P_t(z) - \overline{P_t}] \tag{1.5}$$

Aggregating this micro supply function for the economy as a whole yields the aggregate supply function Y_t. If the density function of the microshock z is $\phi(z)$ with mean zero, we obtain

$$Y_t = \int y_t(z)\varphi(z)dz = \int \theta\gamma[P_t(z) - \overline{P_t}]\varphi(z)dz = \theta\gamma[P_t - \overline{P_t}] \tag{1.6}$$

This aggregate supply function is the core of Lucas' (1972, 73) rational expectations model. As we saw above, this model assumes that micro shocks z are independent of macro shocks (P shocks) and cancel each other out around the macro average. In the model, the distribution of micro shocks z is the same for all economic agents, and each economic agent shares the same distribution, taking it as a constraint for its respective optimization. As we explain in the next section, this assumption is incorrect.

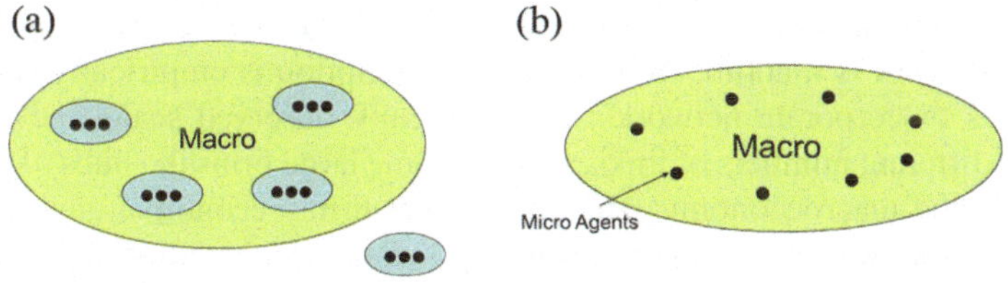

Fig. 1.1 Microeconomic Agents and Macroeconomy

(a) Real economy
(b) Unrealistic assumption in the standard macroeconomic model

1.4 Economic Agent Optimization

In the Lucas model, the distribution of microshocks is the same for all economic agents. Economic agents differ only to the extent that realizations of stochastic variables drawn from the same distribution are different. In reality, however, the micro-constraints faced by economic agents are all different, and there is no one common distribution that serves as a constraint for all agents. Plainly, GAFA and small vegetable merchants do not share the same probability distribution of "microshocks" in each period. In reality, the probability distribution of microshocks differs across economic agents. Moreover, it is likely to be path dependent.

Put another way, in the actual economy, individual economic agents act in different microcosmos, as shown in Fig. 1.1a. However, standard microfounded macroeconomic models such as Lucas all presume a single common macro constraint for all the micro agents, as described in Fig. 1.1b. A typical common macro constraint is the distribution of relevant stochastic variables shared by all economic agents. The Lucas model, labor search models, and recent mean field games are all set up this way. Walras' general equilibrium theory takes a price vector as a common macro constraint shared by all consumers and firms.

Formally, empirical analyses of the network structure of firms demonstrate that this fundamental assumption of modern microfounded macroeconomics is not borne out by fact. Aoyama, et al. (2010), for example, analyzing more than a million Japanese firms, shows that the corporate network is *scale-free*. This means that the firm-size distribution, particularly for large firms, becomes a power distribution. This fact concerning the firm-size distribution itself has been well known for a long time in relation to the *Zipf law*. However, the significance of this fact for microfoundations of macroeconomics is *not* well understood.

Recent network analyses[6] have demonstrated that power distribution is generated by a *scale-free network*, not by a *random network*. In random networks, firm or "node" in the terminology of network analysis is symmetric. It corresponds exactly to Phelps'

[6] Network analysis is a relatively new research field, but has been now firmly established. In the development of this field, the landmarks are Barabási (2016) and Watts (2003).

islands; the Lucas model and labor search models, therefore, tacitly assume that the corporate network is random. However, this assumption is empirically rejected by the analyses of corporate networks. In empirically observed scale-free networks, firms have different numbers of links, making some large firms the hubs of corporate networks in the macroeconomy. This means that firms' economic environment or constraints under which they optimize are not symmetric, but all differ (Fig. 1.1a).

Non-self-averaging —— Nature of Realistic Micro-shocks

The fact that the firm-size distribution obeys the power law and has a "fat tail" leads us to *nonself-averaging* (Sornette, 2004; Fujiwara, 2020). The concept of *nonself-averaging* is almost unknown in economics. Here, we first explain the concept of nonself-averaging. we then explain that the assumption that micro shocks cancel each other out, which is taken for granted in standard microfounded macroeconomics or "rational expectations" models, is actually quite unrealistic.

Consider a random variable X_n that depends on the size of the system n (in physics, an extensive variable). The sum of the random variables, $x_1\ x_2\ \cdots\ x_n$

$$X_n = \sum_{k=1}^{n} x_k \tag{1.7}$$

is X_n. In macroeconomics, we can think of n as the number of economic agents and X_n as a macro variable such as GDP. When the coefficient of variation of X_n, i.e., the ratio of the standard deviation to the mean

$$CV(X_n) = \frac{\sqrt{(\boldsymbol{Variance\,of\,X_n}}}{(\boldsymbol{mean\,of\,X_n})} \tag{1.8}$$

converges to zero as n becomes large ($n \to \infty$), X_n is said to be self-averaging. If it does not converge to zero, the system under consideration is *nonself-averaging*.

The CV is

$$CV(X_n) = \frac{\sqrt{E(X_n - E(X_n))^2}}{E(X_n)} = \sqrt{\frac{E(X_n^2)}{(E(X_n)} - 1} \tag{1.9}$$

Therefore, if it is self-averaging,

$$P((1-\varepsilon)E(X_n) < X_n < (1+\varepsilon)E(X_n)) \to 1 \quad \text{for any}\,\varepsilon > 0,\ n \to \infty \tag{1.10}$$

holds. That is, in a self-averaging system, if the number n of economic agents is large, the movement of X_n can be represented by the average.

As an example of self-averaging, let us look specifically at the Poisson process. It is commonly used in theoretical models in economics such as Aghion/Howitt

(1992). Let λ be the Poisson parameter in which stochastic "events" such as technological progress or job matching occur. In such a model, it is routinely assumed that all economic agents share λ. If there are n economic agents in the economy, the parameters that characterize the macro Poisson process are λn. The average number of times an event occurs in the economy is λn, and the standard deviation is $\sqrt{\lambda n}$. Therefore, the coefficient of variation is $\sqrt{\lambda n}/\lambda n = 1/\sqrt{\lambda n}$ and converges to zero when $n \to \infty$. In other words, the standard Poisson model, which assumes that the same stochastic process applies to all economic agents, is self-averaging. It is easy to confirm that the model that assumes the same normal distribution for all economic agents is also self-averaging (the coefficient of variation converges to zero on the order of $1/\sqrt{n}$, as in the Poisson model).

In such a model, if the size of the model (the number of economic agents) is large, the macroeconomic movement can be captured by the average as described above. In microfounded macroeconomics, the average is determined by the optimization of "representative economic agents" and the equilibrium conditions of the market.

These standard analytical methods, however, only make sense if the model is self-averaging. If it is nonself-averaging, then the macro variables will have significant variation around the mean even when the number of economic agents is large. Therefore, we cannot expect that the changes in the macro variable are captured by the average. In other words, obtaining information about the average by capturing the behavior of representative economic agents through complex optimization becomes meaningless.

Nonself averaging is by no means a pathological or an exceptional phenomenon. The famous physicist Kadanoff (2000), for example, citing the example of the economic phenomenon of asset price fluctuations, says:

> In statistical physics we distinguish between two kinds of statistical movements produced by events that consist of many steps or parts. A simple movement is described by the term self-averaging, when the individual events, when added together, produce an almost deterministic result. ......
>
> In contrast, all multiplicative stochastic processes produce a second set of movements, called non-self-averaging. When the movement of markets and financial instruments can be described by multiplicative stochastic processes, such instruments will exhibit large, almost unpredictable price changes. The greater the number of steps over time, the greater the price uncertainty" (Kadanoff, 2000, pp. 85-86).

We will explain multiplicative stochastic processes in Chap. 7. Here, I follow Sornette (2004, 370) to show how nonself-averaging arises naturally. A stick of length 1 is folded repeatedly. In the first step, divide a bar of length 1 into two bars of length $1-p$ and p $(0 < p < 1)$. The bar of length $1-p$ is kept as it is, and divide the bar of length p in the same proportion. Then, in the second stage, we obtain two bars of length $p(1-p)$ and p^2. Divide the bar of length p^2 in the same way, and so on. If we continue this process of splitting the bars, we obtain a set of bars with length W_n $(n = 1, 2, \cdots)$ as follows:

$$\begin{aligned} W_1 &= 1 - p \\ W_2 &= (1 - p)p \\ &\vdots \\ W_n &= (1 - p)p^{n-1} \end{aligned} \tag{1.11}$$

Let us assume that the fractional probability p at the time of bar folding is a stochastic variable with density function $f(p)$ ($0 < p < 1$). We extract p from this probability distribution and fix a value of p in the bar-folding process as described above. Suppose that we have performed many similar stick-folding processes. Then, consider the variable Y defined in the following equation.

$$Y = \sum_n W_n^2 = \sum_n (1 - p)^2 p^{2(n-1)} = \frac{1 - p}{1 + p} \tag{1.12}$$

By construction, the sum of the length W_n *of* a stick $\sum W_n$ is always equal to 1. The sum of W_n squared, namely, Y, is a stochastic variable because p *is* a random variable and is a measure of the distribution of broken pieces of a stick ($0 < Y < 1$); If Y is close to one, the large piece whose length is almost one exists, whereas if Y is close to zero, all the pieces are tiny. Now, Y, and therefore, W_n is nonself-averaging.

Another example that is much closer to economics and finance is geometric Brownian motion[7]:

$$dX_t = \mu X_t dt + \sigma X_t dB_t \tag{1.13}$$

where μ and σ are constants, and B is Brownian motion, or the Wiener process. Equation (1.13) is the basis of many models, including the Black/Scholes model in finance. The solution of this equation with the initial value X_0 is

$$X_t = X_0 \exp\left[\left(\mu - \frac{\sigma^2}{2}\right)t + \sigma B_t\right] \tag{1.14}$$

If $0 < \mu < \frac{\sigma^2}{2}$, we observe that almost all the trajectories of X_t stochastically converge to zero exponentially. However, the expected value of X_t, $E(X_t)$ is

$$E(X_t) = X_0 \exp(\mu t) \tag{1.15}$$

Thus, the expected value grows exponentially with the growth rate μ. This is a simple example of ergodicity breaking; almost all the trajectories (*i.e.*, measure one) decreases exponentially, but the ensemble average grows exponentially.

Because $E\left(X_t^2\right)$ is.

[7] I owe this basic example of nonself-averaging to Didier Sornette.

$$E(X_t^2) = X_0 \exp[(2\mu + \sigma^2)t], \tag{1.16}$$

the coefficient of variation

$$\frac{\sqrt{E(X_t^2) - E(X_t)^2}}{E(X_t)} \sim \frac{\sqrt{E(X_t^2)}}{E(X_t)} = \exp\left(\frac{\sigma^2}{2}t\right) \tag{1.17}$$

diverges. Thus, geometric Brownian motion is nonself-averaging. It underlines the fact that nonself-averaging is ubiquitous.

To see how nonself-averaging arises in the economic model, consider a very simple model of economic growth (Aoki & Yoshikawa, 2010). We will discuss economic growth in Chap. 5, and this model is meant to be just an example of nonself-averaging, so do not say that there is no microfoundation!

There are two types of innovation in the process of economic growth: (1) innovation that increases the productivity of existing sectors and (2) innovation that creates a new sector or good. Suppose that Kn sectors have been created by the time altogether n innovations have occurred in the economy. In sector i, innovations that increase productivity in this sector occur n_i times ($I = 1, 2, \ldots, K_n$). By definition, if $K_n = k$,

$$n_1 + n_2 + \cdots + n_k = n \tag{1.18}$$

holds. If the n-th innovation creates a new sector (i.e., sector k), then $n_k = 1$.

Y_n is GDP when n innovations have occurred in the economy and is nothing but the sum of the output of each sector y_i:

$$Y_n = \sum_i^{K_n} y_i \tag{1.19}$$

The output y_i of each sector is

$$y_i = \eta\gamma^{n_i} \ (\eta > 0, \ \gamma > 1) \tag{1.20}$$

To facilitate the analysis below, we rewrite Eq. (1.18) as follows.

$$n = \sum_j^n j a_j(n) \tag{1.21}$$

where $a_j(n)$ is the number of sectors in which innovations have occurred j times. We call a vector $a(n)$ composed of $a_j(n)$ the partition vector.[8] Using this partition

[8] See Aoki and Yoshikawa (2007) for the partition vector and the Poisson/Dirichlet distribution described below.

vector, we can rewrite K_n

$$K_n = \sum_j^n a_j(n) \tag{1.22}$$

Using approximation

$$\gamma^{n_i} = \exp(n_i \log \gamma) \cong 1 + \log(\gamma)n_i$$

We can rewrite Eq. (1.20) as follows:

$$y_i = \eta + \eta \log(\gamma)n_i \tag{1.23}$$

From Eqs. (1.18), (1.19), (1.21), (1.22) and (1.23), we obtain

$$Y_n \cong K_n + \beta \sum_j^n j a_j(n) \tag{1.24}$$

While GDP grows with innovation, the impact on growth varies depending on which sector the innovation takes place in.

We must model how innovation occurs. We assume that innovation depends on the two-parameter Poisson/Dirichlet distribution. The two-parameter Poisson/Dirichlet distribution $PD(\alpha, \ \theta)$ is a process that looks like the following. Suppose there are k clusters such that their "size" is n_i $(i = 1, \ 2, \ \cdots, \ k)$ $(n = n_1 + n_2 + \ \cdots \ + n_k)$. In the present model, the cluster "size" corresponds to the number of innovations that occurred in each sector in this model.

Then, the probability p_i of an innovation occurring in an existing sector with size $n_i$is

$$p_i = \frac{n_i - \alpha}{n + \theta} \qquad (\theta + \alpha > 0, \quad 0 < \alpha < 1) \tag{1.25}$$

and the probability of creating a new sector, p, is

$$p = 1 - \sum_i^k \frac{n_i - \alpha}{n + \theta} = \frac{\theta + k\alpha}{n + \theta} \tag{1.26}$$

Note that for the parameter $\alpha = 0$, the model degenerates to a one-parameter *PD* process *PD* (θ). It is clear from (1.26) that when $\alpha > 0$, the probability of the creation of a new sector p is an increasing function of the current number of sectors, k. The number of existing sectors has an externality with respect to the birth of a new sector.

The probability of innovation in this model varies from sector to sector. Therefore, there is no common "representative probability distribution" for all the micro sectors, as the standard microfounded macroeconomics model routinely assumes. The probability of innovation in each sector is not only different but also changes over time. In other words, the probability distribution is *path dependent*. In short, microeconomic agents are not simply different from each other in terms of the realized values from one probability distribution, but they are faced with different stochastic processes.

In this model, which assumes a two-parameter Poisson/Dirichlet process for innovation generation, Y_n or GDP is nonself-averaging (Aoki & Yoshikawa, 2010). This model has a very simple structure and allows for a variety of interpretations. This demonstrates that nonself-averaging naturally arises once we remove the assumption of a "representative probability distribution" of micro shocks.

We can also consider the problem in terms of the transition probability of the Markov model, which is often used as a theoretical model of the economy. Let us represent the transition probability in terms of a tree as shown in Fig. 1.2.

In a one-layered tree such as the one in Fig. 1.2a, the distances between the four states are all equal, and therefore, the four states are symmetric. In term of network analysis corresponds to an empirically rejected random network. On the other hand, in the example of the three-level tree shown in Fig. 1.2b, the distance between states 1 and 2 is much closer than 1 and 3 or 2 and 3. The dynamics of the Markov model with such a large hierarchical structure are not exponential with respect to time but rather lead us to power distribution. This corresponds to the scale-free network observed for the real corporate network structure.

In the standard microfounded model, which assumes that micro shocks cancel each other out, all economic agents are symmetric, so its dynamics correspond to Fig. 1.2a. It is a one-level tree. States 1-4 in Fig. 1.2a are exactly the islands of Phelps, which are represented by representative micro agents. The Lucas rational expectations model that we examined above also has a symmetric tree structure, as shown in Fig. 1.2a.

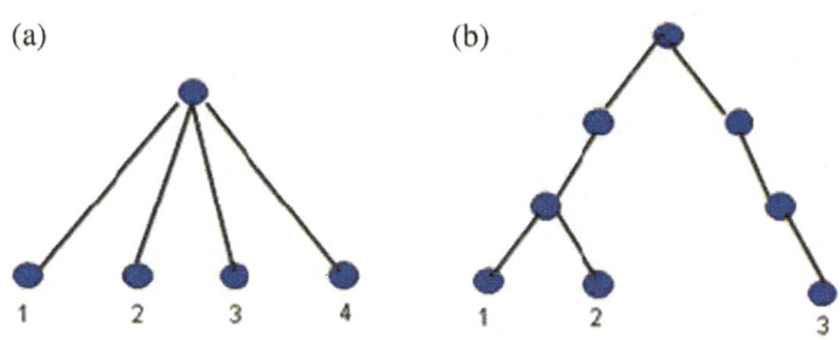

Fig. 1.2 Tree of the transition probability of the Markov model

Note For details, see Aoki and Yoshikawa (2007) Chap. 5

To repeat, under realistic assumptions, including *multiplicative* processes, nonself averaging is expected to occur in general. Under nonself-averaging, there is no representative micro agent, and therefore, ordinary micro foundations become meaningless. Empirical analyses of corporate network structure based on large data have amply demonstrated scale-free networks, meaning that economic agents do not share a common stochastic economic environment. Lucas and Sargent denounced Keynesian economics as "jelly built," but their microfoundations are actually jelly built.

1.5 Rational Expectations

"Rational expectations" have played a decisive role in the major turn of macroeconomics since the 1970s. Arguably, rational expectations represent the fundamental problem of contemporary microfounded macroeconomics.

The papers collected in the collection of Phelps (1970) all assumed "adaptive expectations", and Lucas's paper was no exception. Actually, by then, Muth (1960, 61) proposed "rational expectations" as a hypothesis on expectation formation. Expectations have important, and in some cases, fatal consequences for economic agents, and yet, there are no foundations for expectation formation. This was Muth's motivation; he named "optimal" expectation formation "rational".

Once we are concerned with "optimality", we must make clear the criteria. To do so, it is necessary to specify some economic model that includes variables related to expectations. The models are linear; otherwise, taking the expected values of random variables faces difficulty. At the same time, we need to be clear about how much information the economic agents who shape expectations have. The more information one has, the more accurate one's expectations would naturally be. In other words, even if the posterior performance of expectations is not good, it is not necessarily inconsistent with optimal expectation formation, i.e., rational expectations, if it is due to a lack of information. We will call the set of information that the economic agent has when expectation is formed at time *t the* "information set" and write it as Ω_t.

Now, consider the rational expectations of a variable x based on the information set Ω, where *x is a* stochastic variable of interest. The standard criterion for optimality is to minimize the conditional expected value of the mean square error of x based on information set Ω. In other words, when x^* is "expectation" about x, we minimize

$$E[(x - x^*)^2|\Omega]$$

with respect to x^*, and obtain optimal or "rational expectations". Here, $E[x \mid \Omega]$ is the conditional expected value of x on Ω.

By writing $E[x|\Omega]$ as $\overline{x}$, the above equation can be transformed as follows:

$$
\begin{aligned}
E[(x - x^*)^2|\Omega_t] &= E[(x - \overline{x} + \overline{x} - x^*)^2|\Omega] \\
&= E[(x - \overline{x})^2 + 2(x - \overline{x})(\overline{x} - x^*) + (\overline{x} - x^*)^2|\Omega] \\
&= E[(x - \overline{x})^2|\Omega] + 2E[(x - \overline{x})(\overline{x} - x^*)|\Omega] + E[(\overline{x} - x^*)^2|\Omega]
\end{aligned} \tag{1.27}
$$

The second term on the right-hand side of Eq. (1.27) is zero.

$$
2(\overline{x} - x^*)E[(x - \overline{x})|\Omega] = 2(\overline{x} - x^*)(E[x|\Omega] - \overline{x}) = 0 \tag{1.28}
$$

Therefore, Eq. (1.27) is

$$
E[(\overline{x} - x^*)^2|\Omega] = E[(x - \overline{x})^2|\Omega] + (\overline{x} - x^*)^2 \tag{1.29}
$$

The loss function is the sum of the conditional variance of x and $(\overline{x} - x^*)^2$. The conditional variance of x does not depend on expectation formation but is simply given. The second term $(\bar{x} - x^*)^2$ is minimized when x^* is equal to the conditional expected value of x. Thus, *the optimal expectation that minimizes the conditional expected value of the mean square error, or rational expectations, is the conditional expected value of the variable in question.* This proposition is nothing but a special case of the well-known theorem of orthogonal projection in Hilbert space, which is explained in any standard textbook of functional analysis.

Rational expectations are nothing more than the Ω-conditional expected value of the variable in question, given the model. This is Muth's definition of rational expectations. Muth (1961) considered a partial equilibrium model of an agricultural market. In his rational expectations model, unlike the case of exogenously given adaptive expectations, the rate of adjustment of expectations changes when the price elasticity changes in the model.

It is important to recognize that rational expectations do not have any predetermined rules, as seen from the definition that they are conditional expected values in a given model. In a nonprobabilistic or deterministic model, rational expectation boils down to "perfect foresight", but this is a degenerated exceptional case. In general, they depend on the stochastic process that generates a variable in question. The "adaptive expectations" actually appeared promising in various fields at the time, including Friedman's (1957) permanent income hypothesis. Muth (1960) considered a kind of "inverse problem" of what kind of stochastic process makes adaptive expectations rational.

Here, we take the consumption function as an example. Keynes's consumption function considers consumption to be a function of current-period income. Allegedly, it is incompatible with rational expectations because it assumes the shortsightedness of households.

Suppose that the household's consumption C_t in period t depends on the expected discounted present value of income Y_t from period t into the future (the discount rate is r). That is,

$$C_t = cE\left[\sum_{j=0}^{\infty}\frac{Y_{t+j}}{(1+r)^j}|\Omega_t\right] \quad (0 < \text{c } 1) \tag{1.30}$$

Equation (1.30) assumes rational expectations about future income. For the information set, we have

$$\Omega_t = \{Y_{t-1}, Y_{t-2}, \ldots\ldots\} \tag{1.31}$$

The specific form of the consumption function depends on what stochastic process produces the income Y_t, as explained earlier. For example, if income Y_t is generated by a random walk, then

$$Y_{t+1} = Y_t + \varepsilon_{t+1}, \qquad E\left[\varepsilon_{t+1}|\Omega_t\right] = 0 \tag{1.32}$$

where

$$Y_{t+j} = Y_t + \varepsilon_{t+1} + \cdots + \varepsilon_{t+j}$$

$$E\left[\varepsilon_{t+k}|\Omega_t\right] = 0, \qquad \text{for } k = 1, \ldots, j \tag{1.33}$$

Therefore, we obtain

$$E\left[Y_{t+j}|\Omega_t\right] = Y_t \tag{1.34}$$

In other words, the conditional expected value (rational expectations) of future income is all the current period income. Therefore, by substituting (1.34)) into (1.30), we obtain

$$C_t = c\left(\frac{1+r}{r}\right)Y_t \tag{1.35}$$

for the consumption function. Consumption C_t in the current period is a function of the current period income Y_t.

Needless to say, we do not suggest that a particular type of consumption function expressed as Eq. (1.35) is realistic. We just point out that the hypothesis that current period income has a strong impact on current period consumption does not necessarily assume short-sighted or "irrational" household behavior. If incomes are random walking, this period's income will have a large impact on consumption in this period under "rational expectations".

After the rational expectations hypothesis was proposed by Muth in 1960, ten years in advance of the "rational expectations revolution" by Lucas, Sargent and others, Negishi (1965) was the first to give a full-scale introduction to rational expectations theory. In his book, *Theory of Price and Resource Allocation* (in Japanese) Appendix

II, Negishi introduced rational expectations along with Muth's (1961) and made comments on the possibility of its application to general equilibrium theory.

> However, in the case of general equilibrium model of the market for a large number of goods, as opposed to a single market model, it may be too far-fetched to assume that economic agents process all the information by making full use of a general equilibrium model; it is costly to obtain the information and to process the information obtained. In this case, the rational course of action is to balance the potential benefits and the costs of information in the forecast. To do so, it is reasonable that only relatively easy to obtain information is used, and predictions are made based on simplified approximate model rather than general equilibrium model, or only a part of the model. The formations of economic agents' expectations depend on the difficulty of obtaining information and models they manipulate" (Negishi, 1965, 217).

Negishi's point about the general equilibrium model is basically the problem of applying rational expectations to macroeconomic models. Walras' general equilibrium theory is a theoretical model of the economy as a whole, so it is a macro-model by definition. Rational expectations are the "conditional expected value" of the random variable in question given a stochastic process. When there are multiple economic agents involved (in the macroeconomy, there are necessarily a large number of agents), all economic agents must share a single model. As we explained in the previous section, empirical network analyses have demonstrated that this is not possible for the macroeconomy.

Microeconomic agents are subject to different stochastic constraints. In fact, no consumer relies on a "macromodel" to predict his/her future income when making consumption decisions. Everyone must use a "simplified approximate model" of his/her own income to make projections. It is different for all households because the stochastic processes that produce income for each household are all different.

The ideas that lead to rational expectations in this sense were actually around before Muth. For example, a study of the savings rates of whites and blacks in the United States in the 1950s found that black households had higher savings rates. Initially, this fact was taken as somewhat unexpected, but the subsequent explanation was that black incomes were more volatile than those of whites, i.e., black households were forced to save more in reserve because of greater uncertainty. This is not rational expectations in the strict sense, but it is suido "rational expectations" in the sense that households consume/save by taking into account not only the first-order moment (average) of income but also the second-order moment (variance). More generally, consumption behavior depends on the stochastic process of income. Note, however, that this applies to one household's behavior, not to the economy as a whole.

It is conceivable that the concept of rational expectations could be applied to a single market, which Negishi also refers to as a "simple market," or "partial equilibrium," for "markets. For example, in a single market where a similar pattern repeats from year to year as the market for agricultural and livestock products analyzed by Muth (1961), the rational expectations hypothesis may well be a powerful analytical tool. Alternatively, if governments repeatedly intervene in the foreign exchange market based on mechanical rules over a period of time, it is quite natural for such government intervention behavior to be "folded" into the expectations of private economic agents.

The rational expectations hypothesis comes into play when analyzing a single market where these same patterns repeat. However, in a complex model involving a large number of markets, its applicability is impossible, as Negishi pointed out. After all, *rational expectations are a "micro" logic*. Lucas, Sargent and others have uncritically applied the concept of rational expectations to a macromodel to which it should not be applied and, as a result, have turned macroeconomics into meaningless exercises.

Only a few months after the publication of *General Theory*, Hicks (1936) wrote a review article on the book for the *Economic Journal*. He regarded "the method of expectations" as Keynes's most significant analytical innovation in *General Theory*:

> There thus emerges a peculiar, but very significant, type of analysis. If we assume given, not only the tastes and resources ordinarily assumed given in static theory, but also people's anticipations of the future, it is possible to regard demands and supplies as determined by these tastes, resources and anticipations, and prices as determined by demands and supplies. Once the missing element—anticipations--is added, equilibrium analysis can be used, not only in the remote stationary conditions to which many economists have found themselves driven back, but even in the real world, even in the real world in "disequilibrium."
>
> From the standpoint of pure theory, the use of the method of expectations is perhaps the most revolutionary thing about this book; but Mr. Keynes has other innovations to make, innovations directed towards making the method of anticipations more usable. (Hicks, 1936, 240).

It is an irony that another new "method of expectations," that is, *rational expectations*, would upset Keynes's economics forty years later.

1.6 Walras' General Equilibrium Theory

Before we leave this chapter, it is useful for us to reconsider Walras' general equilibrium theory because it has played such a central role in theoretical economics for the past 100 years. Any theoretical model of economics that economists routinely use today is essentially a variety of Walras' general equilibrium model, although on the surface, it is quite different from Walras' original framework.

In the discussion of Samuelson's neoclassical synthesis, Arrow compared Keynesian economics with Walras' general equilibrium. Microfoundations of Keynesian economics were also considered relative to Walras' general equilibrium theory. Modern microfounded macroeconomics is nothing more than a miniature version of Walras' general equilibrium theory. Lucas's rational expectations model, RBC, and the search model, the so-called "Keynesian" DSGE, are no exception.

The Walrasian concept of equilibrium that forms the basis of modern economics is none other than the Newtonian concept of equilibrium. In fact, a completely different concept of equilibrium —— the statistical equilibrium arose in Walras' time. Unfortunately, for more than one hundred years, economics has not paid serious attention to it. During this time, economists have revered Walras' general equilibrium model as the shine of economic theory. Incidentally, general equilibrium theory traditionally taught in "microeconomics" is a macromodel because it concerns the entire economy consisting of a large number of markets for all goods and services.

In Walras' general equilibrium model, both households and firms maximize their utility and profits, respectively, while facing only the prices of goods and services as constraints. Economists have no doubts about this framework. However, households and firms are actually concerned not only with the prices of goods and services but also with many other constraints, including "quantity". *Price is not everything.*

Herein lies the difference between Marshall and Walras. Marshall considered a "partial equilibrium" in which supply and demand in the market for a single good or service is expressed as a function of price. This demand/supply analysis is perhaps the very first economic model a beginning student of economics encounters. Those who eventually learned Walras' general equilibrium theory may feel as if Marshall's partial equilibrium theory is a childish analytical framework that is merely a special case ($n = 1$) of the former. However, this is a wrong understanding.

Marshall believed that when we focus our attention on a single good or service, we can perform a reasonably meaningful analysis, even if we assume that supply and demand are a function of price only. Conversely, he believed that the same analysis, i.e., expressing demand-supply as a function of prices alone and concentrating attention on changes in prices, cannot be applied to the macroeconomy. In other words, he consciously rejected Walras-type general equilibrium theory.[9] It is not surprising that after the initial correspondence between Marshall and Walras, the relationship between the two parties soon fell into a state of disintegration.

Walras had in mind a well-organized stock exchange and used it as a model for describing the whole economy; he himself states in Chap. 5 "Markets and Competition" of Volume 2 of his main book, *The Essay on Pure Economics*. This was the biggest mistake that Walras made. In the stock market that Walras had in mind, it is certainly true that the people who participate in the market are interested only in the price. However, it is the special nature of financial markets that microeconomic agents' interest is concentrated on price. The real economy where ordinary goods and services are produced/consumed is quite different. Prices play only a limited role as a determinant of supply and demand.

Incidentally, Karl Menger, who along with Walras is considered to be the leader of the marginal revolution, was also well aware that the demand for commodities depended on X-factors other than price. Menger, who emphasized the "selling power of commodities," even said in his main book, *The Principles of National Economy* (Chap. 5, "Price Theory"), that "prices are not responsible for the most basic characteristics of the economic phenomenon of exchange". Menger also corresponded with Walras, but the correspondence eventually faltered as he severely criticized Walras' general equilibrium theory. Although Walras, Menger and Marshall are often lumped

[9] Marshall's (1920) Mathematical Appendix XXI of *Principles of Economics* (pp. 703–704) presents a Marshall-style "general equilibrium model" that includes the factors of production market. Although it is inelegant compared to Walras's theory, Marshall was not passionate about its polish. The reason for this must be that for Marshall, general equilibrium is a useless concept in any way.

together as the "bearers of the marginal revolution", the economics of these three men are fundamentally different.

It was the general equilibrium theory of Walras that has been inherited by modern economics. However, there is a fundamental problem with the general equilibrium theory for analyzing macroeconomic phenomena.

Walras' system is apparently general. That is why many economists have been attracted to it. However, it disqualifies as a framework for analyzing the macroeconomy. In general, equilibrium theory, the number of goods and services, *n,* is arbitrary, so even if the number of goods is 10,000,000, it still stands good. Ten million is fine. It is so general that many economists may be impressed by the "grandeur" of general equilibrium theory. However, as will be explained in detail in Chap. 5, the essence of the capitalist economy we live in is the metabolism, namely, the creation and destruction of goods and services. Many goods and services finish their role in the age and disappear. On the other hand, new goods and services are born every day. Therefore, there is even a problem with the definition of "goods" when we talk about the number of goods, *n*. Flat-panel TVs or soft drink is a term that can mean a hundred different things to consumers. The quality of goods and services called by the same name changes every day. This is the nature of the economy so that a macromodel that simply fixes the number *n* of well-defined goods and services is actually useless.

Unlike raw materials, which can be clearly defined as a single "commodity", the quality of many goods and services cannot be uniquely and clearly defined. In general equilibrium theory, the variable with which the consumer/firm interacts with the outside world is just "price". It is assumed that the demand/supply function is a function of only the price vector, and in the analysis of the stability of the general equilibrium, as seen in Walras' "search process" (*tâtonnement*), changes in prices play a role in regulating supply and demand.

However, the concentration of economic agents' attention on prices is valid only if the quality of "goods" can be uniquely and unambiguously defined and remains unchanged. This assumption holds in the real economy only for the market of "raw materials". Another market in which market participants are interested only in prices is the financial market for stocks, bonds, and foreign exchanges, which we analyze in Chap. 7. Just the opposite is the labor market.

As Walras himself admits in his book, the pioneer of Walras' theory of general equilibrium was Cournot. In his 1838 book, *A Study of the Mathematical Principles of the Theory of Wealth*, Cournot developed a general equilibrium model that even explicitly took into account what we call Walras' law. Cournot considered a general equilibrium model of the purchasing power parity of a number of foreign currencies. In other words, Cournot considered the general equilibrium of a financial market. Since asset markets are precisely those in which supply and demand are adjusted by changes in prices, the Cournot model makes sense.

Walras, on the other hand, reformulated Cournot's model as a macromodel of all goods and services. In doing so, it was the Paris stock exchange, as he himself stated, that inspired Walras about the workings of the market. As we pointed out earlier, this was the greatest mistake that Walras made because the real sector, namely, the markets of the majority of goods and services, including labor, is fundamentally different from the financial markets.

In the stock market, there is only one type of company A stock, and those who buy and sell it have nothing to worry about but the price. The adjustment of supply and demand in the market is basically done through changes in price. This is exactly what Walras' general equilibrium theory presumes. The Paris stock exchange, as Walras saw it, must have worked that way. However, while prices are important information for economic agents who purchase many goods and services, they are not the whole story. We are interested in factors other than price such as "quality," "design," and "brand" when we purchase goods and services.

The accumulation of micro data in recent years, or big data, has made this point clear. Mizuno (2016), for example, summarizes the results of his study with Tsutomu Watanabe. They study tick data of real-time prices and sales of the same TV model on *kakaku.com*, an online price comparison site by several electronics retailers. This analysis of big data reveals that 20% of sellers charge more than 50% higher than the lowest price, and yet consumers are willing to pay that higher price. In online retailing, the so-called "menu costs" for stores to rewrite the prices on display are small. The existence of a price distribution rather than a single equilibrium price demonstrates consumers' preferences for a particular group of stores. This means that there are other important factors than price. They analyzed the same model but still differentiated TVs. In the case of new products and services, price is not the only factor that determines demand.

Another extremely interesting finding reported by Yano (2014) is the analysis of big data based on wearable sensors developed at Hitachi Central Research Laboratory in 2006. In this study, data were used to continuously record the real-time activity of a person who wears a sensor on his/her left wrist. The results, in essence, show that human behavior, which of course includes consumption behavior, is fundamentally driven by "interactions" with our surroundings that are independent of prices.

These research findings must not be considered an interesting but minor "little story" for economics. What the analysis of big data reveals is that preferences change through the direct interactions of multiple economic agents before prices change. These "interactions" themselves have a greater impact on microeconomic behavior than prices. Thus, the general equilibrium theory that focuses on price given preferences and technology, however general it may appear, misses an essential part of our understanding the macroeconomy.

The interactions of micro economic agents are complex. However, for the purpose of macroeconomics, one of the key factors is the "quantity constraint", which is nothing but what Carl Menger called the "selling power of goods".

It is important to recognize that these problems of general equilibrium theory have nothing to do with the flexibility of price. The point is that the real economy is simply *not* the one described by Walras' general equilibrium theory. Hicks wrote about it in his later years. For example, in his posthumous book, *A Manket Theory of Money* (1989), he explains it as follows:

the pre-Keynesians were themselves not good at the problem which is thus thrown back at them. This was chiefly because they had no proper theory of markets. (The Walras model, which has become so popular with modern and especially mathematical economists, requires a form of organization which has rarely if ever existed.) (Hicks, 1989, 2).

Many economists consider the proof of the existence of general equilibrium by Arrow and Debreu (1954) to be the best achievement of postwar economics. Undoubtedly, it contributed to giving economics a "scientific guise". However, insofar as Walras' model itself is meaningless as a macroeconomic model, the proof of the existence of equilibrium within that framework is also actually meaningless.

While Hicks is highly critical of Walras, he had a certain understanding of Marshall. Walras' general equilibrium theory is a macroeconomic model. In contrast, Marshall's "analysis of supply and demand" is not concerned with the economy as a whole but only with the working of one market where supply and demand for one commodity are adjusted through changes in price. During Marshall's lifetime in the 19th century, cotton and many other typical products were still simple and of comparable quality. They were commodities with uniform quality, and such commodities were actively traded in the wholesale market, including futures. Marshall's "partial equilibrium" analysis was a framework for analyzing such a market. Marshall was well aware that a very different approach was necessary for analyzing the macroeconomy.

In any case, times have changed since the beginning of the 20th century.

The typical end-products of manufacturing industry no longer considered of objectively standardizable goods, which could be traded on competitive wholesaler markets: they had become much more various, new products and new varieties being continually devised (Hicks, 1989, 23).

A former Federal Reserve chairman Alan Greenspan made a point of the same issue.

Through the first half of this century the U.S. economy probably could be thought of, at least to a first approximation, in these terms After all, it was not particularly difficult to measure the units of most types of agricultural or mining output, and even much of the output of the manufacturing sector could be measured reasonably well in terms of physical units such as tons of carbon steel or board feet of lumber. Even here, of course, prices differed by types of carbon steel and lumber.

However, as we move into the twenty-first century, the concept of a unit of output is becoming increasingly difficult to craft. Today, an ever growing fraction of overall value added reflects intellectual insight as distinct from physical effort. For a rapidly expanding part of our GDP, the notion of a discrete and well defined unit of output is becoming progressively illusive.

Obviously, such a development is raising exceptionally difficult issues for price measurement (Greenspan, 1997, 3).

For these goods, prices no longer adjust for supply and demand.

The price is one aspect of the offer that is made: there are some characteristics of other aspects which are shared by it (Hicks, 1989, 25).

Hicks described the problem with Walras' general equilibrium theory using the key concepts, *flexible price market* and *fixed price market*. In the real economy, the "flexible price market" shares only a small part of raw materials, and the "fixed price market" is much more important, but Walras' general equilibrium model assumes only the "flexible price market".

We explain how prices are determined in Hicks' "fixed price market" in Chap. 6. For now, we must recognize how complex other conditions than prices are. Unlike objectively observable prices, the micro environment in which households and firms operate is complex and unknowable to a third party who analyzes the macroeconomy; we can never observe micro cosmos in Fig. 1.1a. Thus, it is impossible or meaningless to follow the behavior of representative microeconomic agents in detail to analyze the macroeconomy.

The standard understanding that if prices are flexible, we are led to a Walrasian equilibrium (e.g., Arrow's argument cited earlier) makes sense only in the Walrasian playing field and actually has no meaning for understanding the macroeconomy. If we are to abandon Walras' theory, how can we understand the macroeconomy? Keynes's *General Theory* answered this question. Keynes came up with the "right" answer. However, his explanation was awkward and certainly did not satisfy everyone. Nevertheless, Keynes's *Principle of Effective Demand* provides macroeconomics with the same powerful foundation that the equation of state of gas, Boyle/Charles' law, had played in physics.

Modern macroeconomics, however, has abandoned this principle of effective demand that must be the foundation for analyzing the real economy. The "microfoundations" that became the watchword are nothing more than a false "microfoundation" based on the completely unrealistic assumption that all economic agents share a single possibly stochastic macro constraint for micro-optimization. The approach that attempts to understand the macroeconomy by way of analyzing in detail the behavior of representative consumers and firms is a fallacy.

The next chapter introduces the readers to the method of statistical physics as an alternative to the standard approach. Chapter 3 then uses this method to provide a correct microfoundation for Keynesian economics.

Chapter 2
Statistical Physics and Macroeconomics

When analyzing "micro" economic problems, it is plainly necessary to examine the motives of micro economic agents and their optimization in the context of each problem. Then, as conventional economics has done, we describe the state of affairs with the appropriate concept of "equilibrium". However, this approach makes sense only for micro problems. This was Marshall's approach. Marshall always cautioned "if all other circumstances are equal" (*ceteris paribus*) on the full understanding of the meaning of "partial equilibrium".

Macroeconomics, which analyzes the behavior of the economy as a whole, requires an entirely different approach. As Stiglitz says, macroeconomics is in need of a new paradigm. That is exactly the methodology of statistical physics that we describe in this book. It provides macroeconomics with correct microfoundations. Unfortunately, statistical physics is unknown in economics. This chapter explains the basic concept of statistical physics through a simple example.

In the late nineteenth century, Marshall, Walras, Pareto, and others who created neoclassical economics through the "marginal revolution," all looked to Newton's classical mechanics as the basis for their own theoretical framework. For example, Marshall wrote that "economics took the terms statics and dynamics from physics" (Marshall 1898). Marshall's "physics" here refers to Newtonian mechanics. As pointed out in Chap. 1, Marshall limited the scope of application of such methodology to the "micro" or "partial equilibrium." In contrast, Walras applied it to the macroeconomy. Dreaming of making economics a mathematical "science," Walras wrote: "The revolution in political-economics which we have attempted and even achieved is comparable to the revolutions carried out by Descartes, Lagrange, Maxwell and Helmholtz in geometry, mechanics, physics and physiology."[1] Walras's goal was Lagrange's analytical mechanics, or Newtonian mechanics, which was considered a model of learning in nineteenth-century France.

[1] See Maruyama (2008, 233).

H. Yoshikawa, *Reconstruction of Macroeconomics: Methods of Statistical Physics, and Keynes' Principle of Effective Demand*, Advances in Japanese Business and Economics 3, https://doi.org/10.1007/978-981-19-5264-7_2

However, just as neoclassical economics was being created in the second half of the 19th century, a new approach, later called "statistical mechanics", was born in physics. This is now a method to understand macroscopic phenomena, not only in physics but also in chemistry and biology, but it has never been addressed head-on in the world of economics. Yes, Samuelson left an essay entitled "Gibbs in Economics." However, his remarks in it must represent an atmosphere of the profession: "I have limited tolerance for the perpetual attempts to fabricate for economics concepts of "entropy" imported from the physical sciences or constructed by analogy to Clausius–Boltzmann magnitudes (Samuelson 1990, 256)."

2.1 Variational Principle—Is an Inorganic Particle Fundamentally Different from Us?

The starting point of statistical mechanics is the Boyle-Charles law for gases. This is a combination of Boyle's law (1660), which states that the pressure p of gas inside a container is inversely proportional to the volume of the container V if the temperature is constant, and Charles' law (1787), which states that the volume of gas is proportional to the absolute temperature T under a constant pressure.

$$pV = RT \tag{2.1}$$

where R is a constant that does not depend on the type of gas. It was known before the French Revolution at the end of the 18th century that Eq. (2.1) holds between macroscopic physical quantities such as pressure p, volume V and absolute temperature T.

The problem was how to explain the Boyle-Charles law, a result of "macro" observations, in other words, the "microfoundation" of the theory. It is the kinetic theory of gases. Following Maxwell's work on the velocity distribution of gas molecules (1859), it was established as *statistical mechanics* by Boltzmann at the University of Vienna and subsequently completed by Gibbs at Yale University. In passing, Irving Fisher, a leader of neoclassical economics, was a pupil of Gibbs at Yale University.[2]

The individual molecules that make up the gas move randomly inside the container at a high speed of 644 meters per second; Bolt, who set a world record in the 100-meter track and field race in August 2009, ran at a maximum speed of 12.4 meters per second. All the molecules are moving according to Newton's equation of motion, "mass x acceleration = force".[3] That is, if the position vector is x

---

[2] Why did not the statistical mechanics of Master Gibbs have an impact on Irving Fisher, Gibbs' protégé at Yale University? This is a question of interest in the history of economics.

[3] Modern statistical mechanics is quantum statistical mechanics, and we will also use its results in Chap. 3, but we will not go into the differences between quantum and classical statistical mechanics here. See standard textbooks such as Kittel and Kroemer (1980).

$$m = \frac{d^2x}{dt^2} = f = grad\ V(x) \tag{2.2}$$

where m is the mass of the molecule, f is the force acting on the molecule, and $V(x)$ is the potential function (potential energy) of the position at which the gradient *grad* $V(x)$ becomes a force.

Now, when a molecule at initial time $t = 0$ and $x(0)$ moves to $x(T)$ with $t = T > 0$, the necessary and sufficient condition for x(t) to satisfy the equation of motion (2.2) for $0 \le t \le T$ is that $x(t)$ is the extreme value of the functional

$$F = \int_0^T L(x(t), \dot{x}(t))dt = \int_0^T \left[\frac{m|x|^2}{2} - V(x(t))\right]dt \tag{2.3}$$

The integrated function L of the functional (2.3) is the Lagrangian that represents the difference between the kinetic energy $\frac{m|x|^2}{2}$ and the potential energy $V(x)$. The Euler–Lagrange equation (in economics, it is usually called the Euler equation) giving the extreme value of F is

$$\frac{d}{dt}\left(\frac{\partial L}{\partial \dot{x}}\right) - \frac{\partial L}{\partial x} = 0 \tag{2.4}$$

It is easily confirmed that Eq. (2.4) is nothing but the equation of motion (2.2).

In this way, the motion of inorganic substances, such as molecules of gas, is given as a solution to the variational problem. This is called the "principle of least action' or the "variational principle" (Feynmann 1964 Lectures on Physics, Vol. II, ch. 19). If the behavior of the Ramsey consumer is described by Euler's equation as a solution to the variational problem, it should not be considered a monopoly of human optimization. Many economists naively believe that the complex optimization of human economic behavior is completely different from inorganic physical and chemical phenomena. Humans or life must be certainly different from inorganic matter in many respects. However, for our present purpose, in the sense that the movement of inorganic matter is given as a solution to the variational problem, the movement of inorganic matter is on a par with to human economic behavior. We return to this point later.

2.2 The Method for Macro Analysis

Now, many molecules (on the order of Avogadro's number 6×10^{23}) move around in the container, satisfying the equation of motion (2.2). Theoretically, if we know the initial values (position and velocity vectors) of all the molecules, the subsequent motion must be determined deterministically according to the equation of motion.

The only problem is that it is impossible for humans to know a large number of initial values.

Laplace thought that in the eyes of the demon (the so-called "Laplace demon") who knew all the initial values, everything could be explained deterministically. Of course, in reality, it is impossible to know the initial values of 10^{23} molecules. Even if we knew them, given the interactions between the molecules, we would not be able to solve the differential equation of 6×10^{23} molecules analytically. Fortunately, it is not actually necessary to know the microscopic movements of individual molecules in detail to understand macroscopic phenomena. On the basis of these insights, we purposefully abandon tracking the microscopic details to understand the macroscopic phenomena. This is the starting point and the most basic idea of statistical physics.

For example, the Boyle–Charles law (2.1), which is a macro relation, can be explained as follows. Now, N molecules are moving around in a cube of length L. The mass of each molecule is m, and its velocity v. The actual molecules must be moving around with different speeds in all directions. Here, suppose that the molecules are all moving at the same speed and in parallel to each face of the cube. It is assumed that the molecules continue to move in a straight line without collision. This must be a violent assumption that must be unacceptable to those who are obsessed with details of the motion of individual molecules, i.e., the "micro". However, if we insist on the details of the micro unit, we will not be able to move forward because we will never understand the microscopic movements of each molecule. As we shall see below, we can get richer results about the macro by purposefully discarding the micro details.

The pressure is the force per unit area applied to the wall of the cube when the molecule hits the wall and bounces back. To test it, take the coordinates parallel to each face and consider one face perpendicular to the x-axis. The number of collisions of a molecule moving along the x-axis plus to this surface per unit time is v/L. Since a molecule with velocity v takes L/v time to travel only L distance, it will collide with the surface an average of v/L times per unit time. The momentum changes from mv to $-mv$ when the molecule collides with the surface. Therefore, the force product of one molecule on the wall is $2mv$. Since the cube has six faces, the number of molecules in motion in the direction of interest is N/6.

$$\begin{aligned}\text{(Pressure)} = &\text{(Force product of one molecule on the wall)}\\ &\times \text{(Number of collisions per unit time)}\\ &\times \text{(Number of molecules)}/\text{(Wall area)}.\end{aligned} \tag{2.5}$$

Therefore,

$$p = \frac{(2mv) \times (v/L) \times (N/6)}{L^2} \tag{2.6}$$

holds. Note that $V = L^3$, and we obtain

$$pV = \frac{mv^2 N}{3} = \frac{2}{3}\left(\frac{mv^2}{2}\right)N \tag{2.7}$$

The average kinetic energy $mv^2/2$ of a molecule is proportional to the absolute temperature *T*. That is,

$$\frac{mv^2}{2} = \frac{3}{2}kT \quad (k > 0) \tag{2.8}$$

holds. Therefore, for $R = Nk$, we obtain the Boyle-Charles law $pV = RT$.

Reflecting on the above procedure for explaining macro phenomena, an important lesson for macroeconomics can be learned. One important lesson for macroeconomics is that one must only make wildly simplifying assumptions about the microscopic movements of individual molecules that are plainly impossible in practice. Of course, for a more detailed analysis, the simplifying assumptions described above must be modified accordingly. For example, taking into account the possibility of molecules colliding with each other would obviously modify the above statement, but there is no need to go into it in depth for those of us who think about macroeconomics. The important point is that to understand the "macro" relationship of Boyle-Charles's law at large, it is sufficient to assume very simplified "micro" dynamics.

It is "randomness" that plays an essential role in this theory. As mentioned earlier, it is impossible to follow all the microscopic movements of the vast number of components of a macroscopic system. Each microscopic component must be moving randomly, but we cannot observe it. In addition, even if we knew about it, it would not be of much use to us. Kubo (1952), a famous physicist of statistical mechanics, explained this point as follows.

> What is actually necessary for politics is such a macroscopic state of affairs. What we are interested in is how this macroscopic social condition changes.
>
> This has essentially the same meaning as the problem we face in statistical mechanics. When we look at an object consisting of many particles, the motion of each molecule is not visible to our eyes. Fortunately, we don't feel the need to be bogged down by such things. Statistical mechanics provides us with the macroscopic knowledge that we need from a microscopic standpoint.
>
> There is always something probabilistic at the base of it. In the case of social phenomena as well, when one takes a macroscopic view as I have described above, one should leave the microscopic view of who and what happened, and look at the whole from a statistical standpoint (Kubo 1952, 14).

Entropy

The concept that plays the most important role in statistical physics is entropy. Suppose that a microscopic particle is in state k $(k = 1, \ldots, K)$. Let n_i be the number of particles in state k. Thus, we have the following equation.

$$\sum_{k=1}^{K} n_k = N \tag{2.9}$$

N is the total number of particles. $n = (n_1, n_2, \ldots, n_K)$ represents the state of a macro. The possible number of events when n occurs w_n is nothing more than the number of possibilities in which N (unnamed) balls are randomly thrown into K numbered boxes. Therefore, it is

$$P_n = \frac{1}{K^N} \frac{N!}{\prod_{k=1}^{K} n_k!} \tag{2.10}$$

Since the number of all cases in which N particles are assigned to any of K different states is K^N, the probability P_n *of the* macro state $n = (n_1, n_2, \ldots, n_K)$ is

$$P_n = \frac{1}{K^N} \frac{N!}{\prod_{k=1}^{K} n_k!} \tag{2.11}$$

Under given macro constraints, a state $n = (n_1, n_2, \ldots, n_K)$ such that probability P_n *is* maximized is observed in reality. This is the basic idea of statistical mechanics. This is the same idea as the maximum likelihood method in statistics/econometrics.

We consider the maximization of $\log P_n$ because the maximization of probability P_n and the maximization of $\ln P_n$ are equivalent. Following Sterling's formula, an approximation for a large number x is important.

$$\log x! \cong x \log x - x \tag{2.12}$$

Using this formula, we can rewrite the maximization of $\log P_n$ as follows.

$$S = -\sum_{k=1}^{K} p_k \log p_k \quad \left(p_k = \frac{n_k}{N}\right) \tag{2.13}$$

S is Shannon's entropy and represents the probability of combinatorial cases of randomly made allocations or states n.

When S is maximized with the mean of the system as given, the resulting distribution is exponential. The exponential distribution obtained in this way is called the Boltzmann distribution and plays a fundamental role in statistical physics. In the case of molecular kinetics of gas, the mean energy of the molecule is absolute temperature T. We will explain its application to macroeconomics in Chap. 3.

Now, let me summarize the lessons of statistical physics for macroeconomics. To understand the state and movements of a macro system consisting of many micro agents (hereafter, with economics in mind, we will use the term "micro agents" instead of "molecules" or "particles"), it is meaningless to analyze the detailed behavior of micro agents.

The method of standard microfounded macroeconomics, which assumes representative consumers/firms and considers their behavior as a solution to the variational problem, is meaningless. Today's mainstream macroeconomics, which cries out for "microfoundations", does not actually seriously consider micro behavior. As explained in Chap. 1, if it introduces a stochastic economic environment, it does not sufficiently take into account that the objective functions and constraints of microeconomic agents are all completely heterogeneous and, therefore, presumes representative micro cosmos.

In short, the methodology of optimization of micro agents is not a correct micro foundation for macroeconomics. Instead, we must take into account the probability, or entropy, which is explicitly determined by the number of combinatorial cases.

Note that "randomness" appears in a double sense for macroeconomics: one is the randomness surrounding the individual micro economic agents themselves. The other is that, similar to the motion of gas molecules, even if the behavior of individual microeconomic agents is deterministic, they are so numerous and heterogeneous that they must be regarded as random by those who analyze the macroeconomy as a third party.

2.3 Ramsey Model

The basic framework of modern microfounded macroeconomics is the Ramsey model, although the original intention of Ramsey (1927) was elsewhere. Ramsey consumer optimization is well known (e.g., Blanchard and Fischer 1989). This model is taught as an almost compulsory topic in graduate school macroeconomics curricula around the world. It is used not only in business cycle and growth theories but also in the theory of development and the analysis of financial crisis. The model is so universally used that it is instructive to consider it in detail.

The Ramsey consumer maximizes the present value of the utility obtained from consumption $u(c)$ discounted by the discount rate δ

$$\int_0^\infty u(c_t)e^{-\delta t}dt \tag{2.14}$$

under the following constraint (where n is the population growth rate):

$$\dot{k} = f(k) - nk - c \tag{2.15}$$

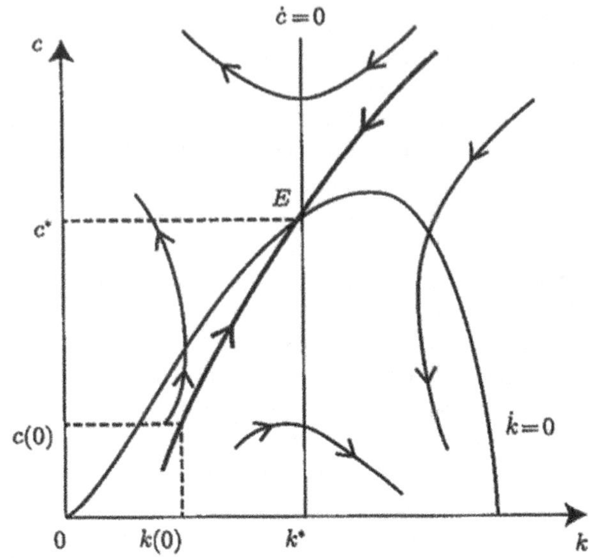

Fig. 2.1 Optimal solution

and given the initial value of per capita capital stock k_0. Here, $f(k)$ is a production function that satisfies conditions $f' > 0$, $f'' < 0$. The consumer's problem is to find the time path of consumption c or capital accumulation that maximizes (2.14).

The qualitative nature of the solution to this optimization problem, obtained using calculus of variational or optimal control, can be seen by the phase diagram (Fig. 2.1). The thick solid line in the figure represents the dynamic optimal path for Ramsey-type consumers. If the initial capital stock is $k(0)$, then the optimal consumption level is $c(0)$. As capital is accumulated, the consumption level of the consumer who became rich will increase. Thus, the pair of the optimal capital stock and consumption moves on the thick solid line in the figure approaching point E (k*, c*).

Some economists may think that humans with a developed brain perform this complex optimization that can be represented as solutions to these variational problems and that they are completely different from the motion of inorganic molecules and particles in this respect. However, as explained in Sect. 2.1, the motion of inorganic particles is also a solution to the problem of calculus of variation (Eqs. (2.3) and (2.4)). In this respect, for the third-party observer, there is no essential difference between the motion of inorganic particles and purposeful human behavior.

Now, the Ramsey model is familiar to all macroeconomists, even to graduate students who have completed a basic course in macroeconomics. There appears to be a large gap between the Ramsey model and the discrete choice model used shortly. The Ramsey model provides "accurate" information about the movements of k and c, as shown in Fig. 2.1. In contrast, the discrete choice model appears to assume too naive and ad hoc behavior of microeconomic agents.

However, the optimal path of c and k (the thick, solid line in Fig. 2.1) truly describes the dynamics of the macroeconomy only if two critical assumptions are met. The first premise is that the macroeconomy consisting of 10^7 consumers can be understood by analyzing the behavior of representative consumers. The second premise is that shock to the representative consumer is absent. In reality, however, neither condition is satisfied.

For example, individual consumers who optimize are constantly subjected to various *micro* shocks. Unemployed consumers will face a worsening of $f(k)$ or income constraints. The utility function $u(c)$ of a sick consumer changes when he/she must go to see a doctor. More than million people die in Japan, and those households hold funerals. The average cost of funerals is 1.89 million yen (2014), which is a large expense for the average household. This also corresponds theoretically to a stochastic disturbance of the utility function, $u(c)$.

It is extremely important to recognize that these shocks are idyosincratic *micro* shocks. They are *not* a drawing from the same probability distribution that all the consumers share. There is no knowledge of the nature of micro shocks for the outside observer.

The fact that consumers are subjected to these micro shocks, such as unemployment and illness, is not inconsistent with the assumption that consumers are "rational" and dynamically optimizing. However, the Ramsey-type consumer in the stochastic economic environment must rethink the Ramsey problem under new conditions every period. In other words, once one accepts the undeniable fact that consumers are subjected to constant shocks to income and preferences, only $c(0)$ is relevant in the Ramsey model, not the whole bold line in Fig. 2.1. In the next period, we must solve the Ramsey problem again under new conditions. As mentioned above, in the next period, in general, both $f(k)$ and $u(c)$ functions themselves change. The optimal path changes correspondingly. In this way, the optimal level of consumption c (and therefore investment k) corresponding to the capital stock in the period is determined on the new optimal path each period.

Thus, the optimal path in a stochastic environment is not the path monotonously approaching point E, as shown in Fig. 2.1. This is the zigzag path (point 1→2→3), as shown in Fig. 2.2. Because the zigzag is determined by the microshocks to which the consumer is subjected, it is not known to a third party what kind of zigzag it is.

What we have considered above is the behavior of a single consumer. As we have seen above, there are 6×10^7 households in Japan, and their behaviors are all in different zigzags. We who analyze the macroeconomy as a third party cannot know these behaviors. Some economists may think that micro zigzags cancel each other out so that if we take the average, we will return to the normal Ramsey consumer's optimal consumption path (Fig. 2.1). However, as already explained in Chap. 1, this is nothing but a declaration. Such an assumption is not justified. Microheterogeneity is not as convenient and simple.

Mean Field Game

An easy approach based on the Ramsey model, which superficially accommodates the heterogeneity of micro agents, is the recent *mean field game* (Achdou et al. 2014). There are many (a continuum of) Ramsey consumers who maximize lifetime utility, (2.14). The constraint is essentially the same as (2.15) but is now

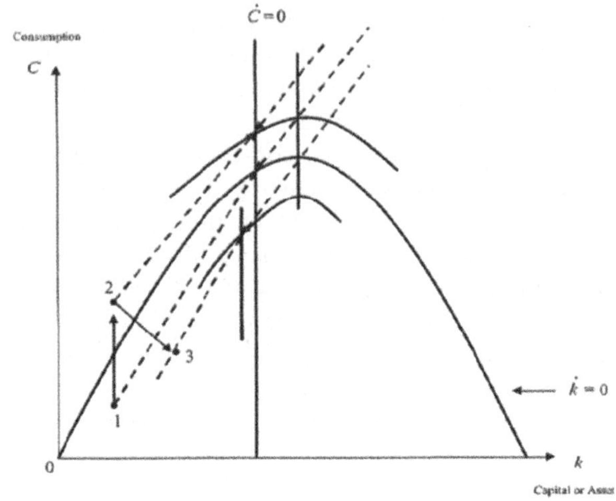

Fig. 2.2 Optimal consumption path in stochastic environment

$$da_t = (Z_t + r(t)a_t - C_t)dt \tag{2.16}$$

$$dZ_t = \mu(Z_t)dt + \sigma(Z_t)dW_t \tag{2.17}$$

Here, a_t is asset or wealth, and Z_t is noncapital income. $r(t)$ is the interest rate, and therefore, $r(t)a_t$ is capital income. The interest rate r is determined in the market and is common to all consumers. In contrast, the noncapital income Z is stochastic, and the realization of Z differs across consumers. Equation (2.17) is the stochastic differential equation (diffusion equation) of Z, where μ and σ are the drift and diffusion terms, respectively, both of which depend on the level of Z. This law of motion of Z is common to all consumers.

In equilibrium, the stochastic version of the Bellman equation, which corresponds to the Euler equation in the standard deterministic model, must hold. It guarantees that the Ramsey consumer maximizes life-time utility under stochastic constraints (2.16) and (2.17). We also obtain the density function g of Z and a:

$$\int g(a, Z)dadZ = 1 \tag{2.18}$$

Consumers differ in terms of Z and a, and we have a stationary density function in equilibrium. On the surface, as in the Lucas rational expectation model, consumers are heterogeneous. However, all the consumers share the same constraints, namely, Eqs. (2.16) and (2.17). The mean field game is another example of a representative stochastic micro environment. In Chap. 1, we explained that this fundamental assumption is not valid.

Krusell and Smith (1998) attempt to defend the representative agents model based on this assumption. In their model, the Ramsey consumer is either employed or unemployed. It is taken as a "micro shock". There is also a stochastic shock to aggregate productivity: there are two possible aggregate states, namely, good and bad. The numbers of agents who are unemployed in two possible aggregate states are assumed to be given and constant. The aggregate shock follows a first-order Markov model determined by transition probabilities. This Markov model is shared by all the consumers and taken by all as a common constraint.

In this model, income and wealth differ across consumers; the title of the paper is "income and wealth heterogeneity in the macroeconomy." However, the authors find that in the stationary stochastic equilibrium, the behavior of the macroeconomic aggregates can be almost perfectly described using only the mean of the wealth distribution. This result is not surprising because their model is a stochastic representative agent model.

2.4 Markov Model of the Business Cycle—An Example

We cannot observe the optimal behavior of micro economic agents at all. This is the starting point for the methodology of statistical physics. Specifically, there is no other suitable method for modeling a 6×10^7 zigzag (which, as already mentioned, is the solution to the Ramsey problem under stochastic economic conditions) than a stochastic process based on the discrete choice model. In the following, we consider a jump Markov process.[4]

An example we consider is Aoki's (1998) model of business cycles. As mentioned earlier, the RBC, starting with Kydland and Prescott (1982), is representative of microfounded macroeconomics. The counterpart of the RBC, a Markov business cycle model, is an excellent model for understanding the implications of the statistical physics method. The business cycle will be discussed in detail in Chap. 4, but here we will consider it as an example for understanding the concept of statistical physics.

Now, the "cycle" is by no means sinusoidal, with boom-bust symmetry (Neftci 1984). The typical pattern is long periods of economic expansion (boom) and short periods of recession (recession). For example, according to the reference dates (peaks and troughs) for Japan from October 1951 to November 2012 (2nd to 15th postwar cycles), the average length of the expansion is 36.2 months, whereas the average length of the recession is 16.1 months.

How do we explain the "asymmetry" in the length of these boom-bust periods? This is the question. In modern microfounded macroeconomics such as real business cycle theory, the candidate to explain the asymmetry of macroeconomic fluctuations must be the preference of representative consumers. Namely, the key parameters that determine the shape of the utility function, such as the elasticity of intertemporal substitution, are expected to explain the asymmetric pattern found in macroeconomic fluctuations.

[4] An explanation of jump Markov process can be found in Chap. 2 of Aoki and Yoshikawa (2007). Van Kampen (1992) is an established textbook for applications.

However, there is no reality in explaining macro phenomena in terms of representative utility function. In the case of the Japanese economy, for example, the global financial crisis that began with the Lehman shock in 2008 caused exports to fall, which led to a serious recession. In 2019, the trade friction between the U.S. and China also led to a decline in the economy. In 2020, the global economy was hit hard by COVID-19. In each episode, where exactly is the connection of the macroeconomy to the utility function of representative consumer?

Now, Aoki (1998) starts from the self-evident fact that the macroeconomy consists of a number of heterogeneous economic agents. In the Japanese economy, the number of households is approximately 60 million, and the number of firms, depending on the definition of firm, is in the millions, so the numbers of households and firms are on the order of 10^7 and 10^6, respectively. This is much smaller than the Avogadro number 6×10^{23} in the natural sciences but large enough for statistical physics methods to be effective.

There are N firms in the macroeconomy. Let each firm either produce at a "high" level y^* or at a low level y ($y^* > y > 0$). If the number of firms producing at a high level of y^* is n, then the output GDP of the economy as a whole, Y, is expressed as

$$Y = ny^* + (N - n)y \qquad (n = 0, 1, \ldots N) \tag{2.19}$$

In this model, firms choose one of two choices: y^*, y. That is, the model is a binary choice model. Many theoretical economists must be reluctant to use such a model for theoretical analysis. Indeed, if one considers the binary choice model literally as it is, it has a very limited scope of application. For example, one might choose to vote for one of the two major parties in election, or one might choose to take a train or a bus as a means of transportation between the two spots. In fact, these issues have been analyzed empirically by MacFaddern and others using econometrics based on the binary choice model. As a result, the binary choice model has a well-established image of a model that should be applied to the choice between two options. Note, however, that for our purpose, it is not at all necessary that the choice of economic agents is a binary choice (y^* and y). This assumption is made just for the sake of simplicity; it is sufficient if the choices are finite (for example, $y_1, \ldots, y_K$; K is an arbitrary natural number).

Now, the share of firms producing at high production level y^* is represented by x.

$$x = \frac{n}{N} \qquad (n = 0, 1, \ldots, N) \tag{2.20}$$

Using x, the GDP expressed in Eq. (2.14) can be rewritten as follows:

$$Y = N\left[xy^* + (1 - x)y\right] \tag{2.21}$$

If N is large, x can be considered a real number ($0 \leq x \leq 1$). Y varies between Ny and Ny^* as x varies between 0 and 1.

x follows a jump Markov process. As mentioned above, individual economic agents (firms) perform dynamic optimization (maximizing the discounted present value of profits or stock price). However, the only way to describe the behavior of a large number (10^6) of firms from a third party's perspective is a stochastic model. At each point in time, some firms lower the level of production as a result of the firm's optimal decision that is not known to the third party. Conversely, other firms increase the level of production. In the model we consider here, the production levels are $y*$ and y ($y* > y$), but as mentioned earlier, this assumption is made for simplicity, and there can be any number of production levels as long as they are countable.

The jump Markov model is completed by determining the *transition probability* of (one) firm changing its production level from each state to the other. Let η_1 be the instantaneous transition rate (transition probability) at which one firm[5] producing at a low production level y moves to a higher production level $y*$. η_1 is equivalent to the birth rate of a woman bearing a baby. The rate at which one firm changes its production from y to $y*$ in the economy as a whole is naturally proportional to the number of firms currently producing at a low level y. This is the same as the probability that a society as a whole will have one more child at a given moment is proportional to the number of women in that society. Therefore, the transition rate r from y to $y*$ is obtained by multiplying η_1 by the number of firms producing at a low level, $N(1-x)$

$$r = N(1 - x)\eta_1(x) \tag{2.22}$$

Similarly, the transition rate at which a firm changes its production from $y*$ to y is

$$l = Nx\eta_2(x) \tag{2.23}$$

η_2 represents the transition rate at which one firm currently producing at $y*$ changes production to y.

In (2.22) and (2.23), the transition rates of one firm η_1 and η_2 depend on the share of the firms currently producing at $y*$, x, which corresponds to GDP, Y. The fact that η_1 and η_2 are not constant but depend on x means that the (optimal) decision of the firm depends on Y. For example, suppose that η_1 (x) is an increasing function of x and η_2 (x) is a decreasing function of x. In this case, the higher the level of GDP is, the more firms currently producing at a low level tend to change their production to a higher level. In other words, when GDP is high, firms become more optimistic than otherwise. Of course, some firms would reduce production even when GDP is high, namely, even in a boom. A firm facing a "structural" decline in demand for the goods it produces must reduce production even in a boom. Some other firms may be forced to scale back production despite a good economy because of ongoing labor disputes or because of a fire in a factory. Conversely, some firms would increase their production levels during a recession when Y is low. This may be the case when a new

[5] Since it is a continuous time model, we can ignore the possibility that two or more firms change their production levels within an extremely short Δt time.

product shows explosive sales. As we have repeatedly stated, it is just impossible for a third party to know the micro situations surrounding individual firms. It can only be modeled as stochastic.

Under these assumptions, $\eta_1'(x) > 0$ and $\eta_2'(x) < 0$ hold. Keynes's principle of effective demand is naturally compatible with these assumptions. In fact, Diamond's (1982) search model, which is meant to provide the principle of effective demand with a microfoundation, can be interpreted as microfoundations of $\eta_1(x)$ and $\eta_2(x)$.[6] Of course, in some cases, the opposite may hold, namely, $\eta_1'(x) < 0$ and $\eta_2'(x) > 0$.

Two transition rates η_1, η_2 must, by definition, sum to 1. In general, η_1 and η_2 satisfying this condition can be written as follows.

$$\eta_1(x) = X^{-1}e^{\beta g(x)} \quad (\beta > 0) \tag{2.24}$$

$$\eta_2(x) = 1 - \eta_1(x) = X^{-1}e^{-\beta g(x)} \tag{2.25}$$

$$X = e^{\beta g(x)} + e^{-\beta g(x)} \tag{2.26}$$

Equations (2.24), (2.25) and (2.26) may seem abrupt at first glance. In fact, however, they are quite natural when considering the decision-making of economic agents in uncertain economic environments. β before $g(x)$ in (2.24) and (2.25) is a parameter that expresses the degree of uncertainty. As the uncertainty increases, β decreases. Conversely, when the uncertainty is small, β is large. See Aoki and Yoshikawa (2007, 60–66) for further explanation. For the moment, it suffices to confirm that if the function $g(x)$ on the shoulder of the exponential function is an increasing function of x, i.e., $g'(x) > 0$, then $\eta_1'(x) > 0$ and $\eta_2'(x) < 0$. Conversely, if $g(x)$ is a decreasing function of x, i.e., $g'(x) < 0$, then $\eta_1'(x) < 0$ and $\eta_2'(x) > 0$. Because X defined in (2.26) appears in the denominators of (2.24) and (2.25), the sum of $\eta_1(x)$ and $\eta_2(x)$ is guaranteed to be 1.

Now, the properties of the transition rates $\eta_1(x)$ and $\eta_2(x)$ explained above correspond one-to-one to the properties of $g(x)$ via (2.24) and (2.25): the larger the value of $g(x)$, the greater the share of firms that think it is more advantageous to change from y to y^*, i.e., to be "bullish" (and vice versa). If $g(x)$ is an increasing function of x, i.e., $g'(x) > 0$, then the number of firms that are bullish tends to increase further when Y is large, and the number of firms that are bearish tends to increase when Y is small. This is an "unstable" case. Conversely, if $g(x)$ is a decreasing function of x, $g'(x) < 0$, this is a "stable" case in which many firms become cautious and reduce production during a boom, and when a recession progresses (x decreases), more firms become aware of the need to "bottom out" and expand their production.

Note that at the zero point $\overline{x}$ of the function $g(x)$, i.e., x satisfying $g(\overline{x}) = 0$, $g = 0$ so that from (2.24) and (2.25), $\eta_1(x) = \eta_2(x) = 1/2$. Under Y, which corresponds to $\overline{x}$, the tendency to expand and contract production is just fifty-fifty.

[6] Diamond's (1982) search model can be represented by a jump Markov model (Aoki and Shirai 2000).

Two examples of functions $g(x)$ are shown in Fig. 2.3. In Fig. 2.3a, $g(x)$ has only one zero point $\overline{x}$ when $g'(x) < 0$. In Fig. 2.3b, $g(x)$ has three zero points A, B, and C. A and C are unstable, whereas B is a stable point. Now, let us check the meaning of $g(x)$ again. $g(x)$ expresses the strength of the tendency of individual firms to find it advantageous to expand their production from y to y^* as a result of respective optimization. If individual optimization is not stochastic, the description of macro phenomena necessarily becomes stochastic, as explained repeatedly, and is represented by the transition rate $g(x)$. It reflects "heterogeneous" optimal behavior. If firms are not heterogeneous, then it makes no sense to consider the ratio x of firms producing at y^* as a probability in the first place. The optimal behavior of a "representative" firm, by definition, must be either y^* or y.

Now, when the macroeconomic environment changes, the function $g(x)$ changes. For example, if macroeconomic policies such as fiscal and monetary policies are changed, the $g(x)$ function itself shifts. When an expansionary macroeconomic policy, such as a decrease in interest rates or an increase in fiscal spending, is invoked, the $g(x)$ function is expected to shift upward (Fig. 2.4a). This is because, as a result of expansionary macroeconomic policy, more firms will find it advantageous to expand output from y to y^* than before, even if GDP or x is the same. As shown in Fig. 2.4a, the zero point of $g(x)$ shifts to the right. Even though investment increases when interest rates are lowered, it does not mean that all firms will increase their investment. Some firms will reduce their investment as interest rates fall. The model we are considering here explicitly takes this possibility into account. The result of expansionary macro policy is not that all the firms produce at y^* but that in the transition

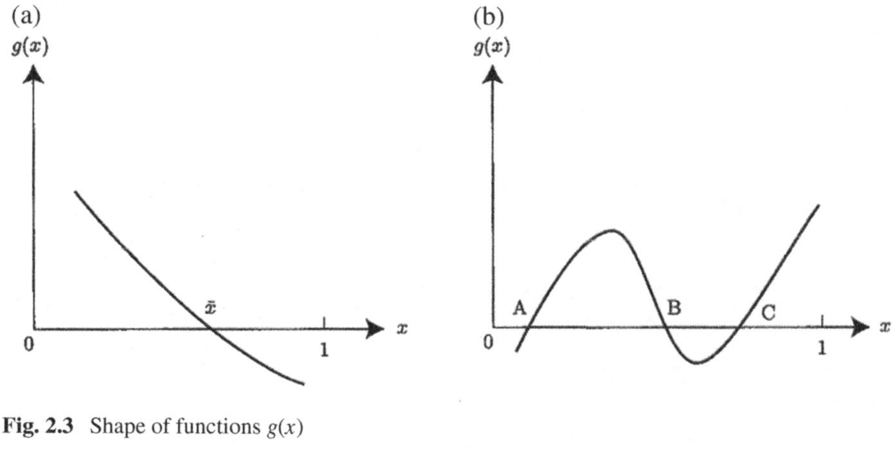

Fig. 2.3 Shape of functions $g(x)$

(a) $g(x)$ has only one zero point $\bar{x}$
(b) $g(x)$ has three zero points A, B, and C

rate from y to y^* under a given x increases. Conversely, if the macroeconomic environment changes in the negative direction, the entire $g(x)$ function shifts downward, and the zero point shifts to the left (Fig. 2.4b).

Let us consider the movements of GDP or Y in this model. Since this is a stochastic model, Y is not unambiguously determined. We consider the probability $P(x, t)$ that the share of a firm producing at a high production level y^* is x at time t. Since x and Y one to one correspond to each other, $P(x, t)$ is essentially the probability that GDP is at the level Y determined by Eq. (2.21) at time t. The master equation to express how $P(x, t)$ changes with time is the so-called Chapman/Kolmogorov equation. Readers interested in this equation should refer to a textbook on stochastic processes or Aoki and Yoshikawa (2007, 93–96). Here, we simply summarize the conclusions.

The movement of the expected value ϕ_t of the stochastic variable x (which also naturally varies over time) is shown by the following ordinary differential equation:

$$\frac{d\phi}{dt} = \phi_t \eta_2(\phi_t) - (1 - \phi_t)\eta_1(\phi_t) \tag{2.27}$$

η_1 and η_2 on the right-hand side are the two transition rates (2.24) and (2.25). The behavior of a random variable x can be known through the differential Eq. (2.27), which represents the movement of the expected value.

We seek the steady-state value ϕ^* of ϕ_t. It is the value of ϕ that makes the right side of the differential Eq. (2.27) zero and is the solution to the following equation.

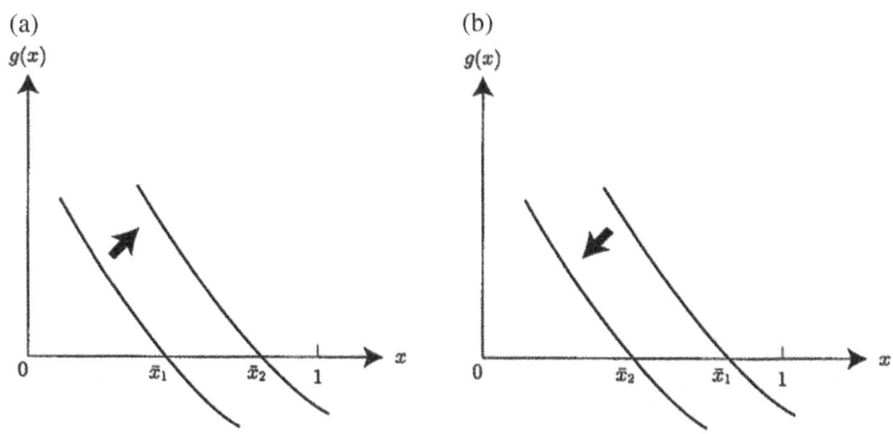

Fig. 2.4 Shift of the $g(x)$ function

(a) Upward shift
(b) Downward shift

$$\frac{\eta_1(\phi)}{\eta_2(\phi)} = \frac{\phi}{1-\phi} \tag{2.28}$$

Equation (2.28) can be rewritten using the transition rate η_1, η_2 definitions (2.24) and (2.25) as follows.

$$2\beta g(\phi) = \log\left(\frac{\phi}{1-\phi}\right) \tag{2.29}$$

As briefly explained earlier, β is a parameter that expresses the magnitude of uncertainty. If the uncertainty faced by economic agents is of little concern, i.e., if β is very large, then Eq. (2.29) is

$$g(\phi) = 0 \tag{2.30}$$

The steady-state value ϕ^* of the expected value ϕ of x *is the* zero point of $g(x)$, i.e., the solution of $g(x) = 0$. However, if the uncertainty cannot be ignored (i.e., β takes a finite value instead of infinity), then the steady state ϕ^* is the solution of Eq. (2.29), not of $g(x) = 0$.

The random variable x fluctuates stochastically around this ϕ^*. In other words, ϕ^* tells us where the "center" of variation of a random variable x (or Y) is. Let us examine the solution of (2.29) slightly more closely. To do so, we use $g(x)$ and β to define a "potential function" $U(x)$ as follows:

$$U(x) = -2\int_0^x g(u)du - \frac{1}{\beta}H(x) \tag{2.31}$$

The function $H(x)$ on the right-hand side is Shannon's entropy defined by

$$H(x) = -x\log x - (1-x)\log(1-x) \tag{2.32}$$

The entropy $H(x)$, which plays an important role in the following analysis, represents the number in the combinatorial case, as explained earlier. In our model, N firms produce at either y^* or y. $H(x)$ is nothing more than the log of the binomial coefficient ${}_NC_n$, namely, the log of the number of firms of N that produce at y^* among N firms. The approximation to $\log_N C_n$ is given by Stirling's formula:

$$\log N! \cong N(\log N - 1) \tag{2.33}$$

Applying (2.33), we obtain

$$\log {}_NC_n = NH(x) \tag{2.34}$$

$H(x)$ represents the number of combinatorial cases in which many economic agents stochastically make binary choices. $H(x)$ is not other than the entropy S explained in (2.13); (2.32) is the case where $K = 2$ in (2.13). The Boltzmann distribution used in Chap. 3 is obtained by maximizing $H(x)$.

Conventional economics has completely ignored $H(x)$, the combinatorial aspect of the problem because it has been implicitly assumed that we, third parties, can accurately observe micro behavior.

In reality, however, we have no idea what the 10^6 or 10^7 micro agents that make up the macroeconomy are doing. The randomness of the micro shocks does not cancel each other out. In that case, as we shall see below, the behavior of x depends crucially on $H(x)$. Therefore, to understand the macroeconomic phenomenon (in this case, the fluctuation of Y), it is wrong to ignore $H(x)$.

Differentiating $H(x)$ with respect to x, we obtain

$$H'(x) = \log\left(\frac{1-x}{x}\right) \tag{2.35}$$

Given (2.35), we know that the extremes of the potential function $U(x)$ defined by Eq. (2.31), i.e.,

$$\begin{aligned} U'(\phi) &= -\,2g(\phi) - \frac{1}{\beta}H'(\phi) \\ &= -2g(\phi) + \frac{1}{\beta}\log\left(\frac{\phi}{1-\phi}\right) = 0 \end{aligned} \tag{2.36}$$

Then, we see that the solution of (2.36) is equivalent to the solution of (2.29). In other words, the steady-state value of the expected value ϕ of a random variable x that we want to know is nothing more than the extreme value (actually the minimum value) of the potential function $U(x)$. The potential function (2.31) is, therefore, a useful function for intuitively understanding the stationary solution of the expected value. Since the potential function depends on $H(x)$ as described in (2.31), we cannot ignore $H(x)$ when understanding macroscopic phenomena (behavior of x or Y). We can ignore the entropy $H(x)$ only in the unrealistic case where β is infinite, i.e., there is no uncertainty at all. In this case, the extreme value of $U(x)$ does not depend on $H(x)$ but is simply the solution of $g(x) = 0$.

In general, Eq. (2.36), i.e., $U'(\phi) = 0$, is a nonlinear function with respect to ϕ and therefore has multiple solutions. That is, it is exceptional for $U(x)$ to have a single extreme value, as shown in Fig. 2.5a. Even if $g(x)$, which determines the transition rate, is a monotonic function, as shown in Fig. 2.3a, it is still quite possible for $U(x)$ to have more than one solution with $U'(\phi) = 0$ due to the existence of a nonlinear term (the second term on the right-hand side of (2.36)) arising from entropy. Aoki and Yoshikawa (1996) analyze the case where the potential function has two minima, as shown in Fig. 2.5b.

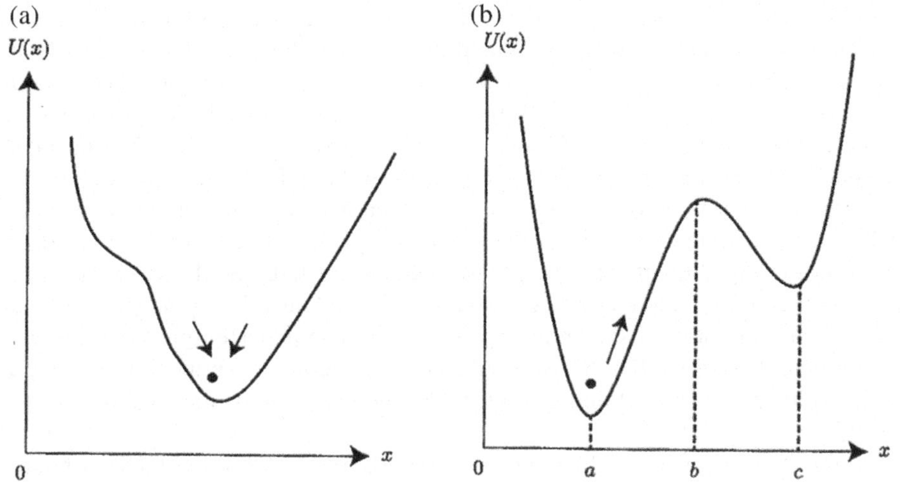

Fig. 2.5 Potential functions

(a) Single extreme value
(b) Two extreme values

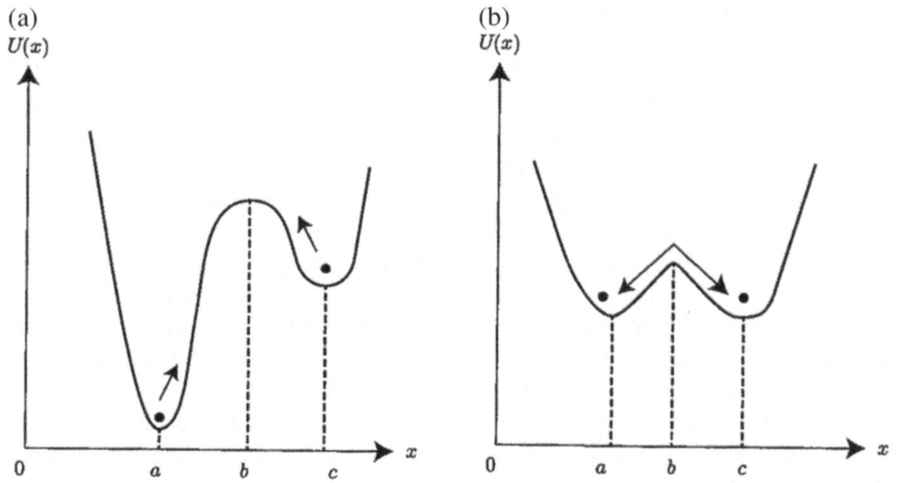

Fig. 2.6 Potential with two extremes

(a) Asymmetric extremes
(b) Symmetric extremes

The stochastic motion of x (and therefore Y) under a potential function with two minima can be understood with the help of Fig. 2.6. The state of x is represented by the black ball in Fig. 2.6. The black ball fluctuates stochastically, but as seen intuitively from the figure, the ball fluctuates most time around stable extremes (a and c in the figure); between the two extremes a and c, there is a hill of unstable extreme b, but over time, the ball moves over the hill of b from a to c, and *vice versa*. The transitions from a to c and from c to a occur stochastically, of course. We can calculate the *mean passage time* of the transitions from a to c or c to a on average. As shown in Fig. 2.6b, if the depths of the valleys of a and c are the same, the mean passage times from a to c and from c to a are almost equal, but if the valley of a is deeper than that of c, as shown in Fig. 2.6a, the mean passage time from a to c is longer than that from c to a. These results are quite natural from the figure. Fig. 2.7 is the result of the simulation by Aoki (1996), which confirms the analytical result nicely.

In this model, point c in the figure corresponds to a high level of GDP, or boom, whereas point a corresponds to a recession. As pointed out at the beginning of this section, the average length of postwar expansions was 36.2 months, while that of recessions was 16.1 months. This asymmetry in the length of the boom-bust period can be easily explained by a potential function, which is shown in Fig. 2.6a. That is, in Fig. 2.6a, if the trough of c is deeper than the trough of a (contrary to the figure), the pattern of longer boom periods than recessions will emerge, as actually observed.

In accordance with the model just described as an example, we point out two things about the methods of statistical physics. The first is the difference between the statistical physics approach and standard microfounded macroeconomics. The macroeconomic phenomenon of the asymmetric length of booms and busts must be explained by microfounded macroeconomics by way of the shape of the utility function of representative consumers.

In contrast, in the statistical-physical approach, macroeconomic movements have nothing to do with the behavior of microeconomic agents (which are all quite different to begin with). We must think carefully about the meaning of "average". Even if we say that the macro phenomenon is, by definition, the average of many micros, that does not mean that there is a representative micro behavior corresponding to that average. Nor do micro agents in the real economy take macro variables as their micro constraints, although there are some exceptions, such as tax codes. As explained in Chap. 1, microenvironments all differ. Entropy $H(x)$ plays a decisive role in macro phenomena. Macro phenomena cannot be understood in terms of the expansion of the representative micro.

Second, the multiple extremes of the potential function correspond to the "multiple equilibria" in standard economic models. The standard approach first considers the multiple equilibria in the model and then *equilibrium selection*, namely, which equilibrium will be selected. The question of how equilibrium is selected has been analyzed in macroeconomics as "history vs. expectations" (Krugman 1991). Various "solutions" have been proposed in the field of game theory as well. However, these "solutions" based on the "rationality" of microeconomic agents are all more arbitrary

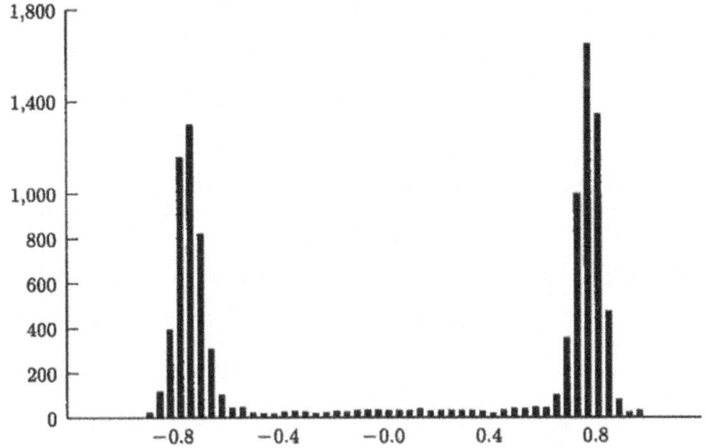

Fig. 2.7 Simulation results. *Source* Aoki and Yoshikawa (2007)

as they become more sophisticated. They may be good exercise for the brain, but they have nothing to do with the real economy.

In the model explained here, even when there are multiple equilibria (i.e., the potential function has multiple extremes), the economy moves stochastically between such multiple equilibria, and therefore, the problem of equilibrium selection does not exist in the first place. As the simulation in Fig. 2.7 shows visually, the economy fluctuates stochastically between two equilibria. We can even calculate the mean passage time required to move from one "equilibrium" to the other. One can think of choosing one equilibrium as a case where the potential trough for that equilibrium is extremely deep, but in any case, the type of potential function depends on entropy completely ignored in macroeconomics. From the viewpoint of statistical physics methodology, the equilibrium selection based on representative agents' optimization itself makes no sense.

The standard methodology of macroeconomics, which examines in detail the optimal behavior of representative micro agents to understand the macro, must be seen by everyone as strange if one broadens his/her horizons from the world of economics and turns to the natural sciences. In natural sciences, whether in physics, chemistry or ecology, the statistical-physical approach is an established method for understanding the macro.

The "microfoundations" of macroeconomics that economists are so confident about are in fact nothing more than a self-congratulatory castle in the sand. Even if consumers and firms engage in sophisticated optimization, the *constraints* perceived by each economic agent in optimizing are so diverse and complex that there is no way for a third party including economists analyzing macroeconomic movements to know them at all. There is no essential difference between macroeconomics and the natural sciences in this respect.

Chapter 3
Stochastic Macroequilibrium: A Microfoundation for Keynesian Economics

3.1 Introduction

The purpose of this chapter is to present a new concept of stochastic macroequilibrium that provides a microfoundation for the Keynesian theory of effective demand. Cyclical changes in aggregate economic activity, namely, quarter-to-quarter or year-to-year changes in real GDP, are basically determined by changes in aggregate demand. This is the central message of Keynes (1936). The fundamental reason why aggregate demand becomes the most important determinant of macroeconomic activity in the short run is that aggregate demand can change swiftly, whereas factor endowment and technology change only slowly.

When aggregate output changes, the utilization of production factors must necessarily change. The most important production factor is labor. The concept of stochastic macroequilibrium clarifies how the labor market works under the Keynesian problem of aggregate demand deficiency. Unemployment and the distribution of labor productivity both depend crucially on aggregate demand. The model is similar to standard search equilibrium in spirit, but it takes a different approach following the method of statistical physics.

The textbook interpretation of Keynesian economics originating with Modigliani (1944) regards it as the economics of inflexible prices/wages. If price and wages were flexible, the economy would be led to the Pareto optimal neoclassical equilibrium. However, whatever the reason, when prices and wages are inflexible, the economy may fall into equilibrium with unemployment. Keynesian economics is meant to analyze such an economy. In this frame of thought, microfoundations for Keynesian economics would provide reasonable explanations for inflexible prices and wages. Toward this goal, a number of studies were performed, once summarized under the heading of "New Keynesian economics" (Mankiw & Romer, 1991). The crux of these studies is to consider optimizing behaviors of the representative household and firm that are compatible with inflexible prices and wages. New Keynesian dynamic stochastic general equilibrium (DSGE) models are built in the same spirit.

H. Yoshikawa, *Reconstruction of Macroeconomics: Methods of Statistical Physics, and Keynes' Principle of Effective Demand*, Advances in Japanese Business and Economics 3, https://doi.org/10.1007/978-981-19-5264-7_3

A different interpretation of Keynesian economics was advanced by Tobin (1993).

> The central Keynesian proposition is not nominal price rigidity but the principle of effective demand (Keynes, 1936, Chap. 3). In the absence of instantaneous and complete market clearing, output and employment are frequently constrained by aggregate demand. In these excess-supply regimes, agents' demands are limited by their inability to sell as much as they would like at prevailing prices. Any failure of price adjustments to keep markets cleared opens the door for quantities to determine quantities, for example real national income to determine consumption demand, as described in Keynes's multiplier calculus. ...
>
> In Keynesian business cycle theory, the shocks generating fluctuations are generally shifts in *real* aggregate demand for goods and services, notably in capital investment (Tobin, 1993, 46–47).

Tobin dubbed his own position an "Old Keynesian view". Certainly, the main message of Keynes (1936) is that *real* demand rather than factor endowment and technology determines the level of aggregate production in the short run simply because the rate of utilization of production factors such as labor and capital endogenously changes in response to changes in real demand. Keynes maintained that this proposition holds true regardless of flexibility of prices and wages; he, in fact, argued that a fall of prices or wages would aggravate, not alleviate the problems facing the economy in deep recession because it may actually lower aggregate demand.

Following Tobin, let us call this proposition the Old Keynesian view. According to the Old Keynesian view, changes in *real* aggregate demand for goods and services generate fluctuations of output. The challenge is then to clarify the market mechanism by which aggregate demand conditions the allocation of production factors in such a way that total output follows changes in real aggregate demand. A decrease in aggregate output is necessarily accompanied by a lower utilization of production factors, and vice versa.

Since the days of Keynes, economists have taken unemployment as the most important sign of possible *underutilization* of labor. However, unemployment is by definition job search, a kind of economic activity of workers, and as such calls for explanation. In addition, unemployment is only a partial indicator of the underutilization of labor in the macroeconomy. As we will explain in Sect. 3.3, the celebrated Okun's law, which relates the unemployment rate to the growth rate of real GDP, demonstrates the significance of *underutilization* of employed labor other than unemployment. In this chapter, we consider not only unemployment but also productivity dispersion in the economy.

To consider Keynes's principle of effective demand, we must obviously depart from the Walrasian general equilibrium as represented by Arrow and Debreu (1954). The most notable example of "non-Walraian economics", which analyzes the labor market in depth, is equilibrium search theory surveyed by its pioneers Rogerson, Shimer, and Wright (2005), Diamond (2011), Mortensen (2011), and Pissarides (2000, 2011). The textbook general equilibrium abstracts itself altogether from the search and matching costs that are always present in the actual markets. By explicitly analyzing search frictions, search theory has succeeded in shedding much light on the workings of the labor market.

However, we find several fundamental problems with the standard search theory. In particular, the theory fails to provide a useful framework for explaining *cyclical* changes in the effective utilization of labor in the macroeconomy (Shimer, 2005). The reason is that although blurred by Poisson modeling, standard theory effectively assumes perfect competition in the product market. No doubt, prices and wages guide economic agents in the market economy. However, quantity constraints also play a crucial role, arguably the most important role in the presence of frictions and uncertainty, which equilibrium search theory emphasizes.

In addition, the optimization of workers and firms in search theory rests on the wrong assumption that all economic agents share common stochastic constraints. As such, contrary to its emphasis on *heterogeneity*, standard search theory is essentially nothing but a stochastic version of the representative agent model. In the macroeconomy, particularly in labor markets, the microenvironment surrounding individual workers and firms is truly diverse, and we cannot explicitly describe it. This means that micro optimization exercises do not make sense because we are unable to set up representative objective functions and constraints. This recognition is the beginning of statistical physics. With the concept of stochastic equilibrium of the macroeconomy based on the method of statistical physics, this chapter clarifies the meaning of the underutilization of labor—productivity dispersion and unemployment—due to demand deficiency.

Section 3.2 points out the limitations of standard search theory. Section 3.3 explains why we must analyze the distribution of labor productivity. After a brief explanation of the concept of equilibrium based on statistical physics in Sect. 3.4, Sect. 3.5 presents a model of stochastic macroequilibrium. The model explains how the distribution of productivity is determined together with unemployment. Section 3.6 explains that stochastic macroequilibrium provides a microfoundation for Keynes's principle of effective demand. It also presents suggestive evidence supporting the model and offers brief concluding remarks.

3.2 Limitations of Search Theory

Search theory starts with the presence of various frictions and accompanying matching costs in market transactions. Once we recognize these problems, we are led to heterogeneity of economic agents and multiple outcomes in equilibrium. In the simplest retail market, for example, with search cost, it would be possible to obtain high and low (more generally multiple) prices for the same good or service in equilibrium. This break with the law of one price is certainly a big step toward reality. Frictions and matching costs are particularly significant in the labor market. The analysis of the labor market has direct implications for macroeconomics. In what follows, we will discuss labor search theory.

In search equilibrium, potentially similar workers and firms experience different economic outcomes. For example, some workers are employed, while others are unemployed, although they are practically similar workers. In this way, search theory well recognizes and even emphasizes the heterogeneity of workers and firms. Despite this recognition, in regard to modeling the behavior of economic agents such as workers and firms, it, in effect, presumes the representative agent in the sense that a *stochastic economic environment is common to all agents*; Workers and firms differ only in terms of the realizations of stochastic variables of interest whose probability distribution is common. Specifically, it is routinely assumed that the job arrival rate, the job separation rate, and probability distributions of wages and productivity are common to all workers and firms. In some models, such as Burdett and Mortensen (1998), the arrival rate is assumed to depend on a worker's current state, namely, either employed or unemployed. However, within each group, the job arrival rate is common to all workers.

The job separation includes layoffs as well as voluntary quits. It makes no sense that all firms and workers face the same job separation rate, particularly the probability of layoffs. White collar and blue collar workers face different risks of layoff. Everybody knows the difference between new expanding industry and old declining industry. In any case, the probability of layoffs depends crucially on the state of demand for the firm's product and, as a result, among other things, on industry, region, and ultimately the firm's performance in the product market. Thus, the probability is far from being common to all firms and workers but is firm-specific and worker specific.

On vacancies and hiring, Davis, Faberman, and Haltiwanger (2013), analyzing monthly data of 16,000 establishments, demonstrate that hires per vacancy significantly differ across firms. More specifically, they find that the greater an expansion of employment of the firm is, the higher the firm's hires per vacancy is. Hires per vacancy is not a parameter common to all firms.

Note that we can always calculate the average job arrival rate and job separation rate, of course. However, the average job arrival rate and the economy-wide wage distribution, which by definition determine the average duration of an unemployment spell, would be relevant only to decisions made by policy makers such as the department/ministry of labor. When individual private economic agents make decisions, they are not relevant because the job arrival rate, the job separation rate, and the wage distribution facing individual workers and firms are all different.

On the assumption that the reservation wage R is the same for all the workers, the standard analysis typically goes as follows. In the equilibrium labor market, we must obtain

$$fu = s(1 - u) \tag{3.1}$$

where u is the unemployment rate, and s and f are the separation rate and the job finding rate, respectively. Equation (3.1) ensures the balance between the inflows and outflows of the unemployment pool.

If λ is the offer arrival rate and $F(w)$ is the cumulative distribution function of wage offers, the job finding rate f is

$$f = \lambda(1 - F(R)). \tag{3.2}$$

From Eqs. (3.1) and (3.2), we obtain

$$u = \frac{s}{s + \lambda(1 - F(R))}. \tag{3.3}$$

As a macro description, Eqs. (3.1) and (3.2) are innocuous. However, when linked to the optimization of micro agents, Eq. (3.2), the standard equation in the literature, in effect, presumes that the offer arrival rate λ, the reservation wage R, and the cumulative distribution function of wage offers F are common to all workers. As we argued above, it is obvious that in reality, λ, R and $F(w)$ differ significantly across workers. It is difficult to imagine that workers of different educational attainments face the same probability distribution of wage offers; it is plainly unrealistic to assume that a youngster working at gas station faces the same probability of getting a well-paid job offer from bank as a graduate of business school. However, in standard search models, the assumption that $F(w)$ is common to all workers is routinely made, and the common $F(w)$ is put into the Bellman equations describing the behaviors of firms and workers. The assumption is simply wrong; $F(w)$ for the ith worker must be $F_i(w)$.

Furthermore, although wages are one of the most important elements in any job offer, workers care not only about wages but also about other factors, such as job quality, tenure, and location. Preferences for these other factors that define a job offer certainly differ widely across workers and are constantly changing over time. Rogerson, Shimer and Wright (2005, 962) say that "although we refer to w as the wage, more generally it could capture some measure of the desirability of the job, depending on benefits, location, prestige, etc." However, this is an illegitimate proposition. All the complexities they refer to simply strengthen the case that we cannot assume that the probability distribution of w, $F(w)$ is common to all the workers. In fact, wage may even be a lexicographically inferior variable to some workers. For example, pregnant female workers might prefer jobs closer to their home at the expense of lower wages. On the part of firms, Davis, Faberman and Haltiwanger (2013) demonstrate the importance of hiring efforts other than vacancies. Their finding underlines the fact that a firm also acts in its own environment. In summary, workers and firms all act in their own different universes (Fig. 1.1a in Chap. 1).

The second problem of the standard search model pertains to the behavior of the firm. Although blurred by standard Poisson modeling, it is routinely assumed that the product market is perfectly competitive in the sense that the *individual* demand curve facing the firm is flat. In Burdett and Mortensen (1998), for example, the flow of revenue generated by employed workers, p is constant, and a firm's steady-state profit given the wage offer w is simply $(p - w)l$, where l is the number of workers of this firm. l depends on the firm's wage setting. This is the standard assumption in the literature, where p is often called "productivity". Given p, the determination of wages plays a crucial role. It is essentially a model of labor shortage in the sense that the firm's output and profits are determined by labor supply at the level of wages the firm offers to workers.

Curiously, standard search theory makes so much effort to consider the determination of wages within firms, taking into account the strategic behavior of rival firms, while at the same time, it leaves the price unexplained under the naïve assumption of perfect competition. Most firms must regard the determination of the prices of their own products as important as (possibly more important than) the determination of wages, taking into account the behavior of their competitors.

Under the assumption of perfect competition in the product market, given a firm's wage offer, whenever a worker comes, the firm is ready to hire him/her, although the worker could turn down the offer or leave the firm when better offers arrive. The level of employment l is determined only by the number of successful matchings. This analytical framework leads us to some awkward conclusions. Postel-Vinay and Robin (2002), for example, interpret the empirically observed decreasing number of workers employed by high productivity firms as a consequence of less recruitment efforts made by high productivity firms than by low productivity firms. However, there is no reasonable reason why high-productivity firms make fewer recruitment efforts. A much more plausible reason is that firms are demand constrained in the product market and that demand for goods and services produced by high productivity firms is limited.

Shimer (2005) also demonstrates that in the standard search model, an increase in the job separation rate raises both the unemployment and vacancy rates. This analytical result is in stark contradiction to the well-known negative relationship between two rates, the Beveridge curve.[1] The strange result is obtained because the standard model assumes perfect competition in the product market and considers only the strategic behavior of firms in the labor market. In an economy constrained by real demand, an increase in the job separation rate is likely to be generated by an increase in layoff, which in turn is caused by a fall in aggregate demand. The unemployment rate rises while firms laying off workers must post fewer vacancy signs and make fewer hiring efforts (Davis, Faberman, & Haltiwanger, 2013).

From our perspective, the most serious problem with standard search theory is the assumption that the product market is perfectly competitive. This assumption is crucial because the firm's demand for labor depends basically on the demand for the firm's products. It is particularly ill-suited for studying *cyclical* changes in the effective utilization of labor. In what follows, following Negishi (1979), we assume that firms are monopolistically competitive in the sense that they face a downward sloping individual demand curve in the product market.

Although the most important factor constraining the firm's demand for labor is demand constraint in the product market, it is absolutely impossible for us to know the individual demand curve facing each firm. The standard assumption in theoretical models of imperfectly competitive markets is that the demand is symmetric, namely, that firms face the same demand condition in equilibrium. This assumption might be justified for the analysis of *an industry* or *a local market* of a single product but is simply wrong for the purpose of studying the macroeconomy. Demand is far from

[1] Most efforts to reconcile standard matching models with the observed data rest on artificial assumptions: Mortensen and Nagypál (2007).

symmetric across firms, and there is no knowing how asymmetric it is in the economy as a whole. Our approach is based on this fact.

In summary, standard search models are built on several unrealistic assumptions. First, the job arrival rate, the job separation rate, and probability distributions of wages and productivity are common to all workers and firms or within a group defined for the purpose of analysis. Some models assume that workers and firms are different in terms of their ability, preferences, and productivity (Burdett & Mortensen, 1998; Postel-Vinay & Robin, 2002; Mueller, 2017). However, the *common* probability distributions of crucial variables are assumed to be utilized by all economic agents; firms strategically determine wages, and workers perform job searches using the same wage distribution and Poisson arrival rates.

We can always find the distribution of any variable of interest for the economy as a whole. The point is that such macro distribution is not relevant to the decisions made by individual economic agents because the "universe" of each economic agent is all different and keep changing (see Fig. 1.1a). In this sense, the macroeconomy is fundamentally different from a local retail market or an industry consisting of a small number of dominant firms. In the case of a local retail market, one may reasonably assume that all the consumers in the small region share the same distribution of prices. However, this assumption is wrong for the economy as a whole. It is a cliché that the labor market is segmented in many ways.

Caballero (2010) makes the same argument with respect to rational expectations:

> While it often makes sense to assume rational expectations for a limited application to isolate a particular mechanism that is distinct from the role of expectations formation, this assumption no longer makes sense once we assemble the whole model. Agents could be fully rational with respect to their local environments and everyday activities, but they are most probably nearly clueless with respect to the statistics about which current macroeconomic models expect them to have full information and rational information (Caballero, 2010, 91).

As we pointed out in Chap. 1, rational expectations are the micro concept that has no relevance to macroeconomics.

Second, the standard assumption that firms are perfectly competitive without any demand constraint in the product market is particularly ill-suited for the analysis of *cyclical* changes in the underutilization of labor. Instead, we must assume that firms face a downward-slowing demand curve. Because we analyze the macroeconomy, we cannot assume that demand is symmetric across firms. Simply, we never know the details of demand constraints facing firms.

The point is *not* that we must *explicitly* introduce all the complexities characterizing labor and product markets into the analytical model. It would simply make the model intractable. Rather, we must fully recognize that it is absolutely impossible to trace the microeconomic behaviors, namely, decision making of workers and firms, in detail. When the problem is micro (e.g., an oligopolistic market dominated by a few firms or sales/purchases of old cars), strategic/optimizing behaviors of economic agents must be explicitly considered. However, the macroeconomy is fundamentally different. In the labor market, microeconomic shocks are indeed *unspecifiable*. Thus, for the purpose of the macroeconomic analysis, sophisticated optimization exercises based on representative agent assumptions do not make sense.

This is actually partly recognized by search theorists themselves. The recognition has led them to introduce the "matching function" into the analysis. The matching function relates the rate of meetings of job seekers and firms to the numbers of the unemployed and job vacancies. The idea behind it is explained by Pissarides (2011) as follows.

> Although there were many attempts to derive an equilibrium wage distribution for markets with search frictions, I took a different approach to labor market equilibrium that could be better described by the term "matching". The idea is that the job search underlying unemployment in the official definitions is not about looking for a good wage, but about looking for a good job match. Moreover, it is not only the worker who is concerned to find a good match, with the firm passively prepared to hire anyone who accepts its wage offer, but the firm is also as concerned with locating a good match before hiring someone.
>
> The foundation for this idea is that each worker has many distinct features, which make her suitable for different kinds of jobs. Job requirements vary across firms too, and employers are not indifferent about the type of worker that they hire, whatever the wage. The process of matching workers to jobs takes time, irrespective of the wage offered by each job. A process whereby both workers and firms search for each other and jointly either accept or reject the match seemed to be closer to reality.
>
> …… It allowed one to study equilibrium models that could incorporate real-world features like differences across workers and jobs, and differences in the institutional structure of labor markets.
>
> The step from a theory of search based on the acceptance of a wage offer to one based on a good match is small but has far-reaching implications for the modeling of the labor market. The reason is that in the case of searching for a good match we can bring in the matching function as a description of the choices available to the worker. The matching function captures many features of frictions in labor markets that are not made explicit. It is a black box, as Barbara Petrongolo and I called it in our 2001 survey, in the same sense that the production function is a black box of technology (Pissarides, 2011, 1093–1094).

What a job seeker is looking for is not simply a good wage but a good job offer that cannot be uniquely defined but differs significantly across workers. It is simply unspecifiable. Pissarides recognizes such "real-world features" as differences across workers and jobs; the "universe" differs across workers and firms. Then, at the same time, he advances the idea of a macro black box. The matching function is certainly a black box not explicitly derived from micro optimization exercises and is, in fact, not a function of any economic variable that directly affects the decisions of individual workers and firms.

Good in spirit, but the matching function is still only a half way in our view. It is based on a kind of common sense in that the number of job matches increases when there are a greater number of both job seekers and vacancies. However, it still abstracts itself from an important aspect of reality. As Okun (1973) emphasizes, the problem of unemployment cannot be reduced only to numbers.

> The evidence presented above confirms that a high-pressure economy generates not only more jobs than does a slack economy, but also a different pattern of employment. It suggests that, in a weak labor market, a poor job is often the best job available, superior at least to the alternative of no job. A high-pressure economy provides people with a chance to climb ladders to better jobs.

> The industry shifts are only one dimension of ladder climbing. Increased upward movements within firms and within industries, and greater geographical flows from lower-income to higher-income regions, are also likely to be significant (Okun, 1973, 234–235).

Unemployment cannot be separated from the qualities of jobs or, more specifically, the distribution of productivity on which we focus here. The dynamics of job ladders are indeed born out by data (Haltiwanger et al., 2018).

Incidentally, there are still attempts to consider heterogeneity within the framework of the standard search model. For example, Mukoyama (2019) criticizes that the standard matching function considers only the aggregate vacancy and, therefore, tacitly presumes the homogeneity of jobs. He introduces heterogeneous jobs in his analysis and replaces V/U with different types of v^i/U, where v^i is the vacancy of type-i jobs and U is the unemployment in the economy as a whole. This v^i/U determines the Poisson rate at which a type-i vacancy finds a worker. However, the heterogeneous job in his model is characterized by the wage for a type-i job, w^i. This is plainly against the spirit of the aggregate matching function. The reason why Pissarides introduced the aggregate matching function is that what job seeker is looking for is not simply a good wage but a "good job" that cannot be uniquely defined. Mukoyama's model emphasizes heterogeneity but actually moves backward from true heterogeneity. In the model, all the workers share the same stochastic environment, and in fact, a worker can apply for all types of jobs simultaneously. Thus, the model is effectively a variant of the representative agent model.

Mueller (2017) using the U.S. microdata from the Current Population Survey for the years 1962–2012 finds that in recessions, the pool of unemployed shifts toward workers with high wages in their previous job and that these shifts are driven by the high cyclicality of separations for high wage workers. The finding is fine. Although he endeavors to reconcile his finding with theory emphasizing the heterogeneity of workers and firms, it remains pure speculation. Differences in the variance of "match-specific shocks", which he argues, are the best candidate for explaining all the relevant facts remain tautological unless it is fully clarified what heterogeneous match-specific shocks exactly mean.

The aggregate matching function does not presume homogeneity of workers and jobs. Rather, it is based on the recognition that matchings of workers and jobs are so diverse that they cannot be described by micro optimization. In considering the problems Pissarides (2011) raised, we face greater complexity and, therefore, need a "greater macro black box" than the standard matching function. The subsequent analysis, in fact, demonstrates that the aggregate matching function is *not* a structurally given function or technology but depends crucially on the level of aggregate demand. This is the motive for the stochastic macroequilibrium we explain in this chapter. Before we proceed to our analysis, we will explain why we must focus on the distribution of productivity.

3.3 Distribution of Productivity as a Measure of Underemployment

The most important implication of the new concept of equilibrium for macroeconomics is that productivities across sectors/firms substantially differ at each moment. In the standard neoclassical theory, production factors are assumed to move across sectors/firms to keep their marginal productivities equal. The existence of adjustment cost modifies this statement only inessentially so long as the adjustment cost is common across agents, as is usually assumed. In equilibrium, marginal productivities must be equated because if they differ, production factors can gain by moving across sectors/firms, and this contradicts the notion of equilibrium. The standard reasoning is that there should be no room for arbitrage in equilibrium.

The equality of productivities across sectors/firms is the fundamental reason why demand does not matter in determining the aggregate output. Demand affects only the composition of outputs. Conversely, the existence of *underemployment*, which is equivalent to the inequality of the value of the marginal product across sectors/firms, means that total output (GDP) can be increased by changing demand.

Suppose that there are n sectors/firms in the economy and that the amount of labor necessary to produce one unit of product in sector i is a_i. Labor is the only input. Taking the good in the least productive sector (sector n) as a numeraire, with p_i as the relative price of the i-th good, we can arrange the order of sectors in such a way that

$$\frac{p_1}{a_1} > \frac{p_2}{a_2} > \cdots > \frac{1}{a_n}. \tag{3.4}$$

Obviously, this order does not depend on the choice of numeraire. p_i/a_i is the value marginal product of labor in terms of good n; sector 1 is the most productive sector, sector 2 the second-most productive sector, and so on. One could interpret sector n as being a low productivity activity such as "household production (housework)" and unemployment (job search). Labor in each sector sums up to a given total labor: $\sum_{i=1}^{n} L_i = L$.

Given the demand for the product of the ith sector D_i, total output in the economy as a whole is

$$Y = \sum_{i=1}^{n-1} p_i D_i + \frac{\left[L - \sum_{i=1}^{n-1} a_i D_i\right]}{a_n} = \left(\frac{L}{a_n}\right) + \sum_{i=1}^{n-1}\left[p_i - \left(\frac{a_i}{a_n}\right)\right] D_i. \tag{3.5}$$

In the neoclassical equilibrium, the value marginal product is equated across sectors: $p_1/a_1 = p_2/a_2 = \cdots = 1/a_n$. Therefore, we obtain

$$Y = \frac{L}{a_n}. \tag{3.6}$$

Total output is independent of demand and depends solely on the endowment of the production factor, L, and technology, a_i.

However, when productivity differs across sectors, i.e., when inequality (3.4) holds, an increase in D_i raises Y. Even a shift of demand increases Y. Y depends on demand. In this case, we say that there is *underemployment* in the economy if there is no unemployment.

It is important to recognize that (involuntary) unemployment is merely a particular form of underemployment and that underemployment, defined as differences in the productivity of production factors, is more important than unemployment for demand to affect total output.

In some models where demand plays a role in the determination of total output, underemployment—or, equivalently, the inequality of productivity across sectors—is assumed. For example, a combination of increasing returns and imperfect competition often leads us to the conclusion that demand matters (Matsuyama, 1995). However, such a combination is nothing but a sufficient condition for underemployment.[2] The structures of the models in which demand plays a role are basically arbitrary, and therefore giving this or that example is of little significance.

For demand to play an essential role in the determination of aggregate output, the inequality of productivity across sectors/activities, or underemployment, is the generic condition. Why and how does the inequality of productivity persist? One might think that differences in productivity across sectors imply unexploited profit opportunities and therefore that they contradict equilibrium. However, heterogeneous economic agents actually have different thresholds for a change in their strategies. Therefore, we cannot suppose that all the economic agents and production factors move instantaneously to the sector with the highest productivity. In fact, as we will see shortly, aggregate demand is the fundamental factor that determines the distribution of productivity.

The celebrated Okun's law demonstrates the significance of underutilization of employed labor other than unemployment or job search. Okun (1962) found that a decline of the unemployment rate by one percent raises the growth rate of real GDP by three percent. Namely, using US quarterly data (1947.2Q–1960.4Q), he analyzed the following regression equation:

$$U_t - U_{t-1} = \alpha - \beta\left(\frac{Y_t - Y_{t-1}}{Y_{t-1}}\right) \tag{3.7}$$

[2] In the model of Matsuyama (1995), for example, two technologies are assumed. One is constant returns to scale in the production of "leisure"; the other is in the production of the "consumption goods", which enjoys increasing returns. In his model, GDP includes not only the consumption goods but also leisure. The government expenditure G affects GDP because the government's propensity to spend on consumption goods is higher than that of households. Thus, an increase in G tilts demand toward the sector that enjoys high productivity. Increasing returns is merely a sufficient condition for demand to affect total output. Increasing returns is necessary for "involuntary unemployment", however; see Weitzman (1982).

where U is the unemployment rate, and Y is the real GDP. He found β to be 0.3. This implies that a decline of the unemployment rate by one percent raises the real growth rate by 3.3%. Such a relationship between the unemployment rate and the real growth rate has been found to be so stable for the US economy that it was subsequently called the "Okun's law".

Now, Okun's law is actually incompatible with the standard neoclassical theory. Suppose that the production function is of the Cobb–Douglas type:

$$Y = K^{1-\alpha} L^{\alpha} \tag{3.8}$$

The theory that the factor price is equal to the marginal product of the respective production factor means that the elasticity of real GDP, Y, with respect to labor, L, is equal to α. Because α is equal to the labor share, it is approximately 2/3. Therefore, given the neoclassical theory, we should expect that a decline of the unemployment rate by one percent—roughly equal to an increase of L by the same magnitude—raises the real growth rate only by 2/3 percent rate than the three percent that Okun found.

Okun (1962) himself explains it by arguing that when the unemployment rate declines, it not only directly increases employment but also generates the following indirect effects: (1) The labor force expands because the size of discouraged workers shrinks. (2) Hours of work per worker increases. Finally, (3) labor productivity increases. In relation to (3), Okun refers to productivity dispersion.

We advance a theory of the distribution of labor productivity and explain how it is fundamentally conditioned by the level of aggregate demand. Before we proceed, it may be interesting to glimpse actual differences in productivity in the economy. The Japanese Ministry of Labor used to compile the country's labor productivity statistics. These results give us very good information on *physical* labor productivity for the same production process in various firms in each industry. Figure 3.1 shows the distribution of labor productivity and wages across firms in the steel industry for 1971. We note that, given a small price dispersion for the same product, the physical labor productivity shown in Fig. 3.1 would reasonably correspond to value productivity and that, to the extent that the production function can be approximated by the Cobb–Douglas type, the *average* productivity shown in the figure would correspond to *marginal* productivity. We can see that the dispersion of productivity is much greater than that of wages. The productivity of the most efficient firm is more than fifteen times that of the least efficient firm.

3.4 Stochastic Macroequilibrium—The Basic Idea

Our vision of the macroeconomy is similar to standard search theory. Workers are always interested in better job opportunities and occasionally change their jobs. Job opportunities facing workers are stochastic depending on vacancy signs posted by firms. Firms, on the other hand, make efforts to recruit and retain the best workers for their purposes.

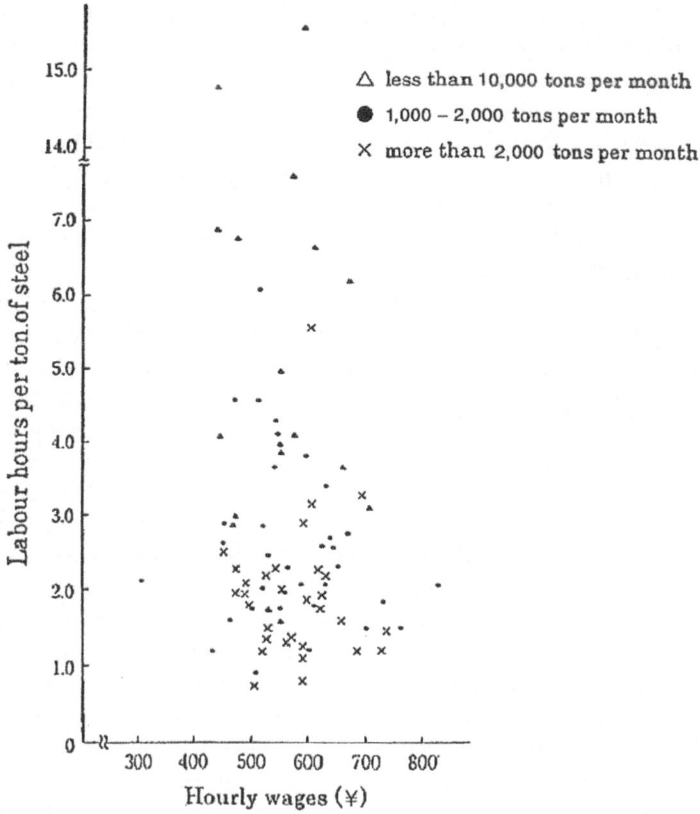

Fig. 3.1 Distribution of Labor Productivity and Wages across Firms in the Japanese Steel Industry, 1971. *Source* Japanese Ministry of Labor, *Survey of Labor Productivity* (1971)

We assume that firms are demand-constrained in the product market. Demand determines the level of production and, as a consequence, demand for labor. In this respect, our model is fundamentally different from standard search theory, which in effect presumes perfect competition in the product market.

Unemployment, a great challenge to any economy, deserves special attention in economic analysis. However, a significant portion of workers actually change their jobs without experiencing any spell of unemployment (Haltiwanger et al., 2018). The importance of on the job search also means that job turnover depends crucially on the distribution of qualities of jobs in the economy. Postel-Vinay and Robin (2002), in fact, analyze on-the-job search explicitly considering workers with different abilities on one hand and firms with different productivities on the other. Their analysis, however, still depends on restrictive assumptions that all the workers face the same job offer distribution independent of their ability and employment status and that the product market is perfectly competitive.

While workers search for suitable jobs, firms also search for suitable workers. A firm's job offer is, of course, conditional on its economic performance. The present analysis focuses on the firm's labor productivity. On the supply side, the firm's labor productivity increases thanks to capital accumulation and technical progress or innovations. However, those job sites with high productivity remain only *potential* unless firms face high enough demand for their products; firms may not post job vacancy signs or even discharge the existing workers when demand is low. Indeed, Davis, Faberman, and Haltiwanger (2013) show that when the firm raises the level of production, it strengthens hiring efforts.

As noted above, we assume that firms are all monopolistically competitive in the sense that they face downward sloping individual demand curves, and the levels of production are determined by demand rather than increasing marginal costs (Solow, 1986). Unlike market-determined prices of raw materials and commodities, prices of most manufactured products and services are made by firms facing downward sloping demand curves (Okun, 1981).

Formally, a most elegant general equilibrium model of monopolistic competition is given by Negishi (1960–61). Furthermore, Sweezy (1939) and Negishi (1979) persuasively argue that when the firm is monopolistically competitive, the individual demand curve is actually not only downward-sloping but must be *kinked at* the current level of output and price. The response of the firm's sales to a change in price is asymmetric because of the asymmetric reactions of rival firms on one hand and the asymmetric reactions of customers on the other. Drèze (1979) also shows that for a risk-averse firm, uncertainty about the price elasticity of demand has an effect equivalent to that of a kinked demand curve with the kink located at the current price and quantity. Therefore, in an economy with uncertainty and frictions, as emphasized by search theory, it is reasonable to expect a kinked individual demand curve facing a monopolistically competitive firm.

The corresponding marginal revenue then becomes discontinuous, as shown in Fig. 3.2. Analytically, firm j ($j = 1, 2, \ldots, K$) maximizes its profit

$$p_j(y_j, \overline{p}_j, \overline{y}_j)y_j - wn_j \tag{3.9}$$

with respect to labor n_j. Here, y_j and p_j are the firm's output and its price, respectively.

For simplicity, we assume a linear production function.

$$y_j = c_j n_j \qquad (c_j > 0) \tag{3.10}$$

where c_j is labor productivity. In fact, if the marginal cost were significantly increasing, there should be a strong incentive for firms facing fluctuations in demand to smooth out production. However, firms actually do not smooth production in response to predictable seasonal cycles of demand. This suggests that the marginal cost is virtually constant; see also Shea (1993).

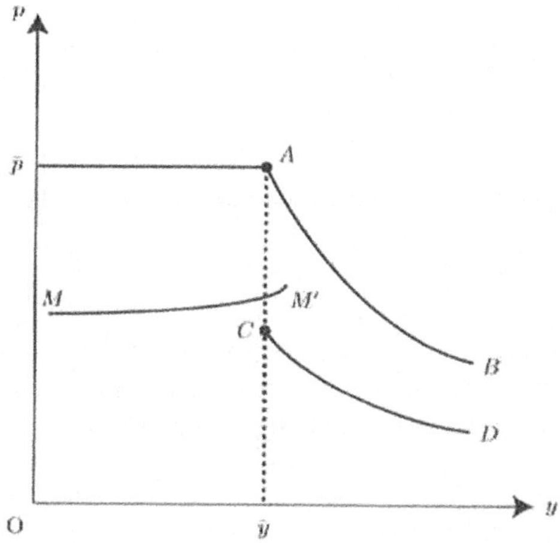

Fig. 3.2 The Kinked individual demand curve facing monopolistically competitive firm. *Source* Negishi (1979)

Note $\overline{p}$ AB = Individual demand curve
$\overline{p}$ A CD = Marginal Revenue
MM′ = Marginal Cost

The price p_j depends not only on y_j but also on the current $\overline{p}_j$ and $\overline{y}_j$. The firm perceives that it is prevented by competition from raising the price above its current level $\overline{p}_j$, while at the same time, it realizes the difficulty of selling a larger quantity of its product without reducing the price below $\overline{p}_j$. Thus, the perceived individual demand curve $p_j(y_j, \overline{p}_j, \overline{y}_j)$ satisfies the following conditions:

$$\overline{p}_j = p_j\left(\overline{y}_j, \overline{p}_j, \overline{y}_j\right) \tag{3.11}$$

$$p_j = \overline{p}_j \quad \text{for} \quad y_j \leq \overline{y}_j \tag{3.12}$$

and

$$\frac{\partial p_j}{\partial y_j} < 0 \quad \text{for} \quad y_j > \overline{y}_j. \tag{3.13}$$

When the firm's expectation of sales is realized, i.e., $y_j = \overline{y}_j$ and $p_j = \overline{p}_j$, the conditions for maximization are

$$p_j(1 - \eta_j)\, c_j \leq w \tag{3.14}$$

and

$$p_j c_j \geq w \tag{3.15}$$

at $(\overline{p}_j, \overline{y}_j)$, where η_j is the right-hand side elasticity of p_j with respect to y_j at $\overline{y}_j$ (Fig. 3.2).

Conditions (3.14) and (3.15) mean that the marginal cost of production w/c_j is no larger than the price and no smaller than the marginal revenue $p_j(1 - \eta_j)$ for any increase in the level of output. It can be shown that any change in the level of sales at the unchanged price, namely, a shift in demand, is completely absorbed in a change in output, leaving the price unchanged because inequality conditions (3.14) and (3.15) remain undisturbed. We must note that this result does not ensue only from the assumption that the firm's individual demand curve is downward-sloping. If there is no kink in the perceived demand curve, an increase in sales at the unchanged price may be followed by an increase in price leaving output and labor employment unchanged. With the kinked individual demand curve, the marginal revenue becomes discontinuous. Under this assumption, a shift in demand is absorbed in all changes in output and labor employment, leaving the price unchanged.

We must also note that given inequality conditions (3.14) and (3.15), a change in wages that disturbs the marginal cost leaves output and labor employment unchanged. This result is in stark contrast to the standard search literature. In equilibrium search theory, wages determined within firms by means of such a mechanism as bargaining crucially affect the job creation rate and accordingly the level of output. This is because labor employment and output are assumed to depend on a firm's strategic behavior of recruitment under the assumption of perfect competition in the product market, which, in turn, depends on wages by way of profitability. In the standard search framework, therefore, the determination of wages plays a crucial role in accounting for volatility of employment or unemployment (Pissarides, 2009). However, it is hard to believe that wages, regardless of how they are determined, have anything to do with a sharp increase in the US unemployment rate from 4.6 to 9.8% during the Great Recession (2007–2010). In the spring of 2020, due to COVID-19, consumption abruptly fell, and the US unemployment rate skyrocketed from 4.4% in March to 14.7% in April. Meanwhile, wages hardly changed.

In the present framework, the way wages are determined is actually of secondary importance for the determination of labor employment and output because within a certain limit, wages do not affect inequality conditions (3.14) and (3.15). Here, output and labor employment are basically determined by real demand facing individual firms in the product market, not by wages. With the perceived kinked individual demand curve, a shift of demand is basically absorbed by a change in output leaving price unchanged. As Tobin (1993) says, the model "opens the door for quantities to determine quantities." In this framework, we can concentrate on quantities (output and labor employment) without explicitly considering price and wages. This is what happened during the Great Recession and the COVID-19 recession.

Changes in wage every moment do not affect price either (Fig. 3.2). However, when increases in wages are nation/industry wide and taken as permanent, price will be raised with little change in output and employment; Point A moves upward parallel with a shift of marginal cost curve MM' in Fig. 3.2. This is mark-up pricing based on *normal* production costs (Godley & Nordhaus, 1972).

Negishi (1979; Chap. 6) shows the existence of a general equilibrium of monopolistically competitive firms facing the kinked individual demand curve. Dreze and Herings (2008) establish the existence of general equilibrium under much more general assumptions. Although equilibrium exists, it is indeterminate because the initial levels of the firms' output and price (the kink point of the individual demand curve, or point P in Fig. 3.2) must be exogenously given. To give such an initial condition amounts to determining the level of the individual demand curve facing each firm.

The present analysis provides a rule for allocating the aggregate demand D to monopolistically competitive firms based on the principle of statistical physics.

$$D = \sum_{j=1}^{K} \overline{y} \tag{3.16}$$

Negishi (1979) and Dreze and Herings (2008) general equilibrium models of monopolistically competitive firms abstract itself from frictions and uncertainty in the labor market. Unlike capital, however, workers either stay at a job, move to another, or search for one as the unemployed. Whenever necessary, firms try to recruit the best workers for their purposes. These are, of course, aspects of the labor market that search theory analyses. We follow its spirit but not its analytical method.

In the present analysis, we assume that firms with higher productivity make more attractive job offers to workers. We do not know how attractive they are to which workers. The assumption simply means that whenever possible, workers move to firms with higher productivity, although we never know particular reasons for such moves. For workers to move to firms with higher productivity, it is necessary that those firms must decide to fill vacant job sites and post a sufficient number of vacancy signs and/or make sufficient hiring efforts. They post such vacancy signs and make hiring efforts only when they face an increase in demand for their products and decide to raise the level of production. It also goes without saying that high-productivity firms keep their existing workers only when they face sufficiently high demand.

The question we ask is what the distribution of employed workers is across firms whose productivities differ. As we argued in the previous section, because microeconomic shocks to both workers and firms are so complex and unspecifiable, optimization exercises based on representative agent assumptions do not help us much. In particular, we never know how the aggregate demand is distributed across monopolistically competitive firms. To repeat, among other things, the job arrival rate, the job separation rate, and the probability distribution of wages (or more generally measure of the desirability of the job) differ across workers and firms.

This recognition is precisely the starting point of the fundamental method of statistical physics. Foley (1994), in his seminal application of this approach to general equilibrium theory, called the idea the "statistical equilibrium theory of markets". At first, one might think that allowing too large a dispersion of individual characteristics leaves so many degrees of freedom that almost anything can happen. However, the methods of statistical physics provide us not only with qualitative results but also with quantitative predictions.

In the present model, the fundamental constraint on the economy as a whole is aggregate demand D. Accordingly, to each firm facing the downward-sloping kinked individual demand curve, the level of demand for its product is the fundamental constraint. The problem is how the aggregate demand D is allocated to these monopolistically competitive firms, namely, the determination of $\overline{y}_j$. Our model provides a solution. The basic idea behind the analysis can be explained with the help of the simplest case. For the moment, we focus on productivity dispersion. We will introduce unemployment into the model in the next section.

Suppose that n_k workers belong to firms whose productivity is $c_k(c_k < c_{k'}$ where $k < k')$. There are K levels of productivity in the economy $(k = 1, 2, \ldots, K)$. The total number of workers N is given.

$$\sum_{k=1}^{K} n_k = N \tag{3.17}$$

A vector $n = (n_1, n_2, \ldots, n_K)$ shows a particular allocation of workers across firms with different productivities. The combinatorial number W_n of obtaining this allocation, n, is equal to that of throwing N balls to K different boxes. It is

$$W_n = \frac{N!}{\prod_{k=1}^{K} n_k!} \tag{3.18}$$

Because the number of all the possible ways to allocate N different balls to K different boxes is K^N, the probability P_n that a particular allocation $n = (n_1, n_2, \ldots, n_K)$ is obtained is

$$P_n = \frac{W_n}{K^N} = \frac{1}{K^N} \frac{N!}{\prod_{k=1}^{K} n_k!}. \tag{3.19}$$

It is the fundamental postulate of statistical physics that the state or the allocation $n = (n_1, n_2, \ldots, n_K)$ that maximizes the probability P_n or (3.19) *under macroconstraints* is to be realized.[3] The idea is similar to maximum likelihood in statistics/econometrics.

Maximizing P_n is equivalent to maximizing $\log P_n$. Applying the Stirling formula for a large number x

[3] To be precise, it is to be realized in the sense of expected value. In physics, variance is normally so small relative to expected value that we practically always observe the expected value.

$$\log x! \cong x \log x - x, \tag{3.20}$$

we find that the maximization of $ln P_n$ is equivalent to that of S.

$$S = -\sum_{k=1}^{K} p_k \log p_k \qquad (p_k = \frac{n_k}{N}) \tag{3.21}$$

S is the Shannon entropy and captures the combinatorial aspect of the problem. It can be easily understood with the help of a simple example. When we throw a pair of dice, the possible sums range from two to twelve. Six is much more likely than twelve simply because the combinatorial number of the former is five whereas that of the latter is only one. Although the combinatorial consideration summarized in the entropy plays a decisive role for the final outcome, that is not the whole story, of course. The qualification "under macroconstraints" is crucial.

The first macroconstraint concerns the labor endowment (3.17). The second macro constraint concerns the effective demand.

$$D = \sum_{k=1}^{K} y_k = \sum_{k=1}^{K} c_k n_k \tag{3.22}$$

Here, aggregate demand D is assumed to be given. In our analysis, we explicitly analyze the allocation of labor $(n_1, n_2, \ldots, n_K)$. The allocation of labor is basically equivalent to the allocation of the aggregate demand to monopolistically competitive firms. Note, however, that for firms, demand is given, whereas employment is the outcome of firms' hiring efforts.

To maximize entropy S *under two macroconstraints* (3.17) and (3.22), we set up the following Lagrangean form L:

$$L = -\sum_{k=1}^{K} \left(\frac{n_k}{N}\right) \log\left(\frac{n_k}{N}\right) + \alpha \left[N - \sum_{k=1}^{K} n_k \right] + \beta \left[D - \sum_{k=1}^{K} c_k n_k \right] \tag{3.23}$$

with two Lagrangean multipliers, α and β. Maximization of this Lagrangian form with respect to n_k leads us to the first-order conditions:

$$\log\left(\frac{n_k}{N}\right) = -1 - \alpha N - \beta N c_k \qquad (k = 1, 2, \ldots, K) \tag{3.24}$$

which are equivalent to

$$\frac{n_k}{N} = \exp[-1 - \alpha N - \beta N c_k]. \tag{3.25}$$

Because n_k/N sums up to one, we obtain

$$\frac{n_k}{N} = \frac{e^{-\beta N c_k}}{\sum_{k=1}^{K} e^{-\beta N c_k}} \tag{3.26}$$

Thus, the number of workers working at firms with productivity c_k is exponentially distributed. It is known as the Boltzmann distribution in physics.

Here arises a crucial difference between economics and physics. In physics, c_k corresponds to the level of energy. Whenever possible, particles tend to move toward the *lowest* energy level. In contrast, in economics, workers always strive for better jobs offered by firms with higher productivity c_k. As a result of optimization under unobservable constraints, workers move to better jobs. In fact, if allowed, all workers would move up to the job sites with the highest productivity, c_K. This situation shown in Fig. 3.3 corresponds to the textbook Pareto optimal equilibrium with no frictions and uncertainty.

However, this state is actually impossible unless the level of aggregate demand D is so high as equal to the maximum level D_{max} defined as $c_k N$. Only when D is equal to D_{max} do demand-constrained firms with the highest productivity face high enough individual demand curves so that they have incentives to employ N workers. On the part of workers, they would be able to find job sites with the highest productivity because firms with c_K try hard to keep their existing workers and, at the same time, post enough vacancy signs and make aggressive hiring efforts.

The maximum level of aggregate demand D_{max} is only imaginary, however. When D is lower than D_{max}, the story is quite different. Some workers—a majority of workers, in fact, must work at job sites with productivity much lower than c_K.

How are workers distributed over job sites with different productivity? It depends on the level of aggregate demand. When D reaches its lowest level,

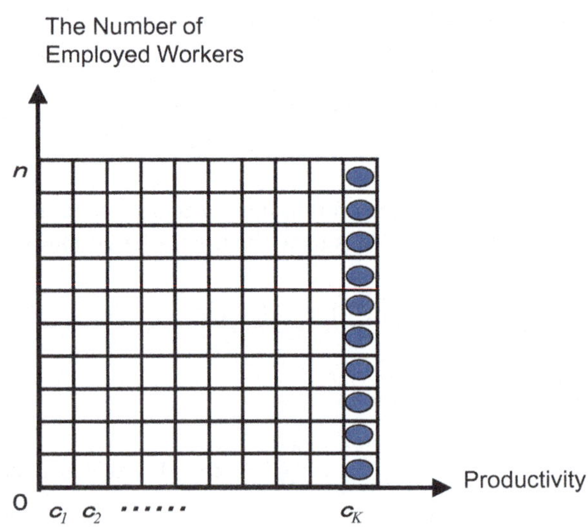

Fig. 3.3 All workers work at job sites with the highest level of productivity under extremely high aggregate demand, D_{max}

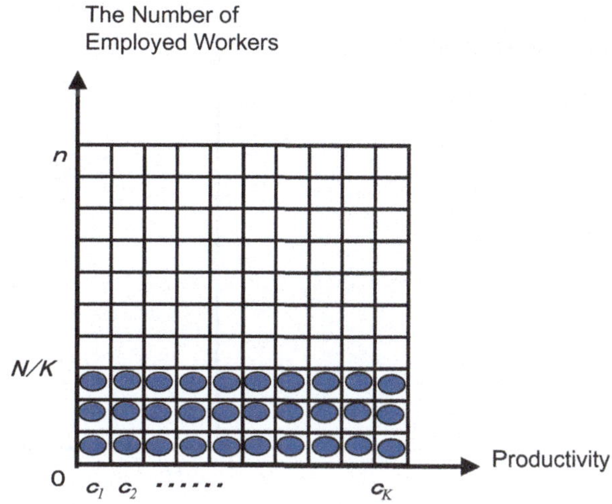

Fig. 3.4 Workers are distributed evenly across all sectors with different levels of productivity under extremely low aggregate demand, D_{min}

D_{min}, workers are distributed evenly across all sectors with different levels of productivity, $c_1, c_2, \ldots, c_K$ (Fig. 3.4). Here, D_{min} is defined as $D_{min} = N(c_1 + c_2 +, \ldots, +c_K)/K$. It is easy to see that the lower the level of D is, the greater the combinatorial number of distribution $(n_1, n_2, \ldots, n_K)$ that satisfies the aggregate demand constraint (3.22) becomes. Note that this combinatorial number basically corresponds to the distribution of demand for goods and services across job sites at monopolistically competitive firms with different levels of productivity.

As explained above, the combinatorial number W_n of a particular allocation $n = (n_1, n_2, \ldots, n_K)$ is basically equivalent to the Shannon entropy, S defined by (3.21). S increases when D decreases. In the extreme case where D is equal to or greater than the maximum level D_{max}, all the workers work at job sites with the highest productivity because whenever possible, workers move to better job sites with higher productivity (Fig. 3.3). In this case, the entropy S becomes zero, its lowest level, because $n_K/N = 1$ and $n_k/N = 0 \quad (k \neq K)$. In the other extreme where aggregate demand is equal to the minimum level D_{min}, we have $n_k = N/K$ (Fig. 3.4), and the entropy S defined by (3.21) becomes lnK, its maximum level.

The relation between the entropy S and the level of aggregate demand D, therefore, resembles the relation shown in Fig. 3.5. In the present model, entropy basically corresponds to the dispersion of job quality in the labor market, which in turn comes from the dispersion of demand facing firms in the product market. Workers face greater uncertainty and friction when the level of aggregate demand is low because the dispersion of job quality becomes larger in that case. Low-productivity firms can attract workers, as high-productivity firms facing more stringent demand constraints make fewer job openings under lower aggregate demand.

At this stage, we can recall that the Lagrangean multiplier β in (3.23) for the aggregate demand constraint is equal to

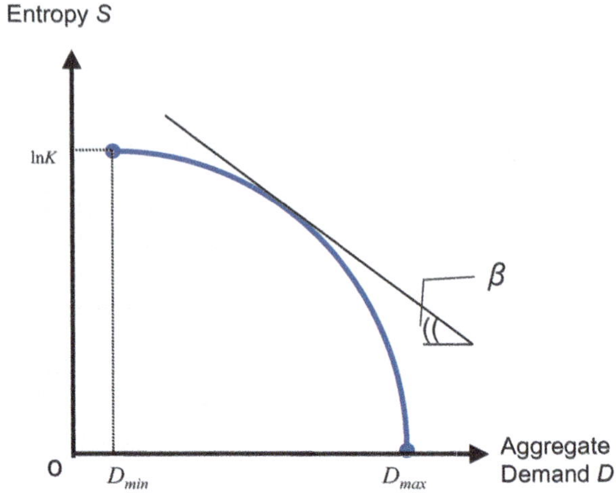

Fig. 3.5 Entropy S and Aggregate Demand D

$$\beta = \frac{\partial L}{\partial D} = \frac{\partial S}{\partial D}. \tag{3.27}$$

β is the slope of the tangent of the curve, as shown in Fig. 3.5, and is therefore negative.[4] Table 3.1 shows how negative β relates to the level of aggregate demand D.

With negative β, the exponential distribution (3.26) resembles the one shown in Fig. 3.6. Whenever possible, workers always move toward job sites with higher productivity. As a result, the distribution is upward-sloping. However, unless the aggregate demand is equal to (or greater than) the maximum level, D_{max}, workers' efforts to reach job sites with the highest productivity c_K must be frustrated because firms with the highest productivity do not need to employ a large number of workers and are less aggressive in recruitment; it becomes harder for workers to find such jobs. As a consequence, workers are distributed over all job sites with different levels of productivity.

The maximization of entropy under the aggregate demand constraint (3.23), in fact, balances two forces. On the one hand, whenever possible, workers move to

Table 3.1 Negative β and aggregate demand

β	-0	……	$-\infty$
Aggregate demand D	Low	……	High

[4] In physics, β is positive in most cases. This difference arises because workers strive for job sites with higher productivity, not the other way round (Iyetomi, 2012)). In physics, β is equal to the inverse of temperature, or more precisely, temperature is defined as the inverse of $\partial S / \partial D$ when S is the entropy and D energy. Thus, negative β means the *negative temperature*. It may sound odd, but the notion of negative temperature is perfectly legitimate in such systems as the one in the present analysis. In particular, the number of possible levels of energy must be finite; see Appendix E of Kittel and Kroemer (1980). In our present case, the level of productivity is plainly finite.

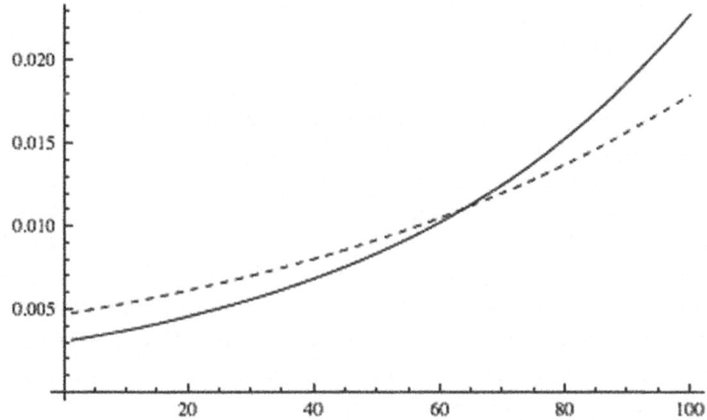

Fig. 3.6 Distribution of labor across sectors with different productivity. *Source* Yoshikawa (2015)

better jobs identified with job sites with higher productivity here. It is the outcome of successful job matching resulting from the worker's search and the firm's recruitment. When the level of aggregate demand is high, this force dominates. However, when D is lower than D_{max}, there are in general a number of different allocations $(n_1, n_2, \ldots, n_K)$ that are consistent with D.

As we explained in the previous section, micro shocks facing both workers and firms are truly unspecifiable. We simply do not know which firms with what productivity face how much demand constraint and need to employ how many workers with what qualifications. We do not know which workers are seeking what kind of jobs with how much productivity, either. Here comes the maximization of entropy. This gives us the distribution $(n_1, n_2, \ldots, n_K)$, which corresponds to the maximum combinatorial number consistent with a given D.

The entropy maximization under the aggregate demand constraint plays, therefore, a role similar to that of the matching function in standard search theory. Note that unlike the standard matching function, which focuses only on the number of jobs, the matching of job quality characterized by productivity plays a central role in the present analysis. The matching function presumes that the number of job matches increases when the numbers of unemployed individuals and vacancies increase. In the present analysis, the matching of high productivity jobs is ultimately conditioned by the level of aggregate demand because high aggregate demand loosens demand constraints facing monopolistically competitive high productivity firms, and *vice versa. Uncertainty and frictions emphasized by standard search theory are not exogenously given but depend crucially on aggregate demand.* In a booming gold-rush town, one does not waste a minute finding a good job! The opposite holds true in a depressed city.

The distribution of workers over job sites with different productivity levels that results from this analysis is an upward-sloping exponential distribution, as shown in Fig. 3.6. In the figure, the bold line ($\beta = -1/5000$) corresponds to high aggregate demand, D and the dotted line ($\beta = -1/7500$) to low aggregate demand. As one should expect, the higher the level of aggregate demand D is, the steeper the distribution becomes, meaning that more workers are mobilized to jobs with higher productivity (Okun, 1973).

It is essential to understand that the present approach does *not* regard economic agents' behaviors as random. Certainly, firms and workers maximize their profits and utilities. The present analysis, in fact, presumes that workers always strive for better jobs characterized by higher productivity. However, firms are demand-constrained facing downward-sloping kinked demand curves. As a result of profit maximization, the optimal level of production is constrained by demand. Unless the level of aggregate demand is extremely high, micro demand constraints become effective. In this case, the entropy matters because we can never know details of micro behaviors. Randomness underneath the entropy maximization comes from the fact that both the objective functions of and constraints facing a large number of economic agents are constantly subject to *unspecifiable* micro shocks. We must recall that the number of households is of order 10^7 or 10^8, and the number of firms, 10^6. Therefore, there is nothing for outside observers, namely, economists analyzing the macroeconomy but to regard a particular allocation *under macroconstraints* as equi-probable. Then, it is most likely that the allocation of the aggregate demand and workers that maximizes the probability P_n or (3.19) *under macroconstraints* is realized.[5]

3.5 The Model

The above analysis shows that the distribution of workers at firms with different productivities depends crucially on the level of aggregate demand. Although the simple model is useful to explain the basic idea, it is too simple to apply to the

[5] This method has been time and again successful in natural sciences when we analyze object comprising many micro elements. Economists might be still skeptical of the validity of the method in economics saying that inorganic atoms and molecules comprising gas are essentially different from optimizing economic agents. Every student of economics knows that behavior of dynamically optimizing economic agent such as the Ramsey consumer is described by the Euler equation for a problem of calculus of variation. On the surface, such a sophisticated economic behavior must look remote from "mechanical" movements of an inorganic particle which only satisfy the law of motion. However, every student of physics knows that the Newtonian law of motion is actually nothing but the Euler equation for a certain variational problem; particles minimizes the energy or the Hamiltonian! It is called the principle of least action: see Chap. 19 of Feynmann (1964)'s *Lectures on Physics*, Vol. II. Therefore, as we explained in Chap. 2, behavior of dynamically optimizing economic agent and motions of inorganic particle are on a par to the extent that they both satisfy the Euler equation for respective variational problem. The method of statistical physics can be usefully applied not because motions of micro units are "mechanical," but because object or system under investigation comprises many micro units individual movements of which we are unable to know. It leads us to a specific exponential distribution (3.26).

empirically observed distribution of labor productivity. In addition, it abstracts itself from unemployment. We introduce unemployment into the model.

Empirical Distribution of Productivity

Figure 3.7a–c and show the distributions of workers at different productivity levels for the Japanese economy ((a) total, (b) manufacturing and (c) nonmanufacturing industries). The data used are the Nikkei Economic Electric Database (NEEDS, http://www.crd-office.net/CRD/english/index.html) and the Credit Risk Database (CRD, http://www.crd-office.net/CRD/english/index.html), which cover more than a million large and medium/small firms for 2007.

The "productivity" here is simply the value added of the firm divided by the number of employed workers, that is, the average labor productivity. Theoretically, we should be interested in unobserved *marginal* productivity, not average productivity. In addition, proper "labor input" must be in terms of work hours or, for that matter, even in terms of work efficiency units rather than the number of workers. For these reasons, the average labor productivity shown in the figure is a crude measure of theoretically meaningful unobserved marginal productivity. However, Aoyama et al. (2010, 38–41) demonstrate that when the average productivity and measurement errors are independent, the distribution of true marginal productivity obeys the power law with the same exponent as that for the measured average productivity. In other words, the distribution is robust with respect to measurement errors in the present case. Incidentally, we also note that the Pareto efficiency of the economy pertains to marginal products of production factors, *not* to total factor productivity (TFP) some economists focus on.

Figure 3.7 drawn on the double logarithm plane broadly shows that (1) the distribution of labor productivity is single-peaked, (2) in the low productivity (left) region, it is upward-sloping exponential, whereas (3) in the high productivity (right) region, it obeys downward-sloping power law (Aoyama et al., 2010). Ikeda and Souma (2009) find a similar distribution of productivity for the U.S. DelliGatti et al. (2008) find power-law tails of productivity distribution for France and Italy. In what follows, we present a model for explaining the broad shape of this empirically observed distribution. The extended model also explains unemployment.

We explained in the previous section that entropy maximization under macro constraints leads us to an exponential distribution. This distribution with negative β can explain the broad pattern of the left-hand side of the distribution shown in Fig. 3.7a–c, namely, an upward-sloping exponential distribution(Iyetomi, 2012). However, we cannot explain the downward-sloping power distribution for high-productivity firms. To explain this, we need to make an additional assumption that the number of potentially available high-productivity jobs is limited and that it decreases as the level of productivity rises.

Dynamics of Potential Job Creation/Destruction

Potential jobs are created by firms by accumulating capital and/or introducing new technologies, particularly new products. On the other hand, they are permanently destroyed by firms' losses of demand for their products. Schumpeterian innovations

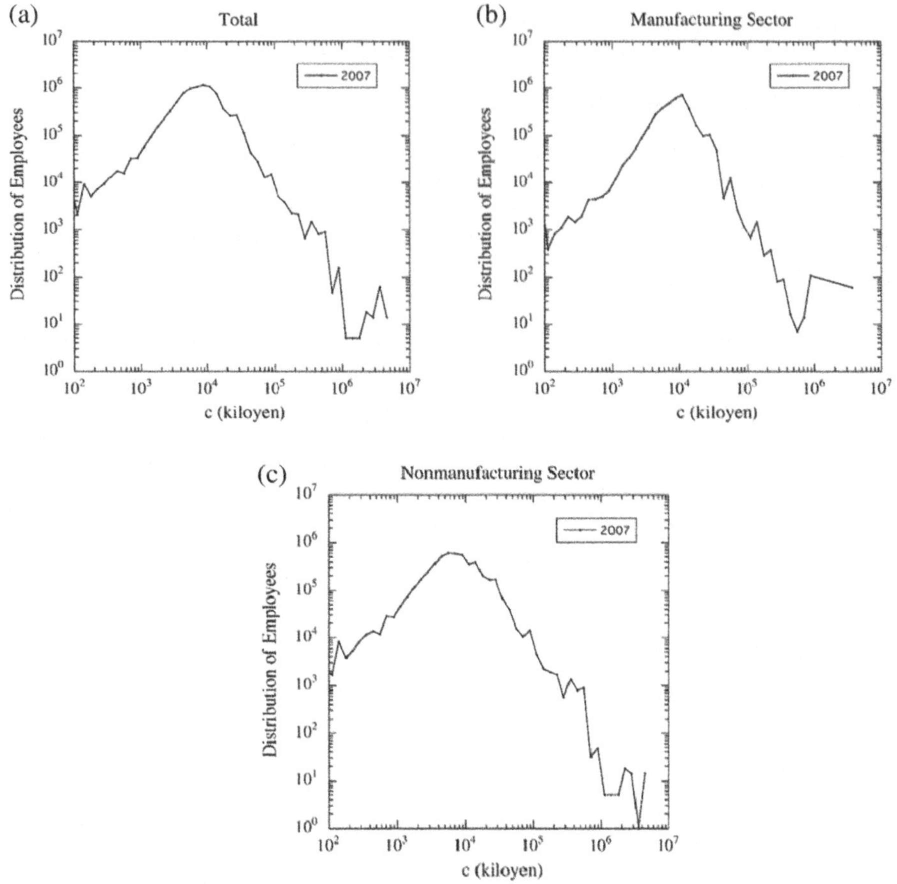

Fig. 3.7 Distribution of Labor Productivity in Japan (2007)

(a) Mean = 1.29×10^4, Standard Deviation = 2.37×10^4
(b) Mean = 1.30×10^4, Standard Deviation = 2.21×10^4
(c) Mean = 1.28×10^4, Standard Deviation = 2.48×10^4

Source Yoshikawa (2015)

by way of creative destruction raise the levels of some potential jobs but at the same time lower the levels of others. In this way, the number of potential jobs with a particular level of productivity keeps changing. They remain *potential* because firms do not attempt to fill all job sites with workers. To fill them, firms either keep the existing workers on the job or post job vacancy signs and make enough hiring efforts, but they are economic decisions and depend crucially on the economic conditions facing the firms. The most important constraint on the firm's employment is the level of demand in the product market. The number of potential job sites, therefore, is not exactly equal to but rather imposes a ceiling on the sum of the number of filled job sites, namely, employment and unfilled jobs.

The statistical theory we will discuss later explains how employment is determined. In this subsection, we first consider the dynamics of *potential* job creation and destruction. Causes of creation/destruction of potential job sites are microshocks and, therefore, unspecifiable. The best way to describe them is the Markov model. Good examples in economics are Champernowne (1950) on income distribution and Ijiri and Simon (1979) on the size distribution of firms. Here, we adapt the model of Marsili and Zhang (1998) to our own purpose. The goal of the analysis is to derive a power-law distribution such as the one for the tail of the empirically observed distribution of labor productivity.[6]

Suppose that there are f_j "potential" jobs with productivity, c_j ($c_j < c_{j'}$, when $j < j'$). In a small time interval dt, the level of productivity c of a potential job site increases with probability $w_+(c)\,dt$ by a small amount, which we can assume is unity without loss of generality. We denote this probability as $w_+(c)$ because w_+ depends on the level of$c$. Likewise, it decreases by a unit with productivity$w_-(c)\,dt$. Thus, $w_+(c)$ and $w_-(c)$ are transition rates for the processes $c \to c+1$ and $c \to c-1$, respectively. As noted above, the productivity of job sites changes for many reasons. It may reflect technical progress or innovations. Given the significant costs of adjusting labor, productivity also changes when demand for a firm's products change. For example, when demand for a firm's products falls, labor productivity necessarily declines. The decline of productivity in this case reflects labor hoarding (Fay & Medoff, 1985). These changes are captured by the transition rates, $w_+(c)$ and $w_-(c)$, in the model.

We also assume that a new job site is born with a unit of productivity with probability pdt. On the other hand, a job site with productivity $c = 1$ will cease to exist if c falls to zero. Thus, the probability of exit is $w_-(c = 1)\,dt$. A set of the transition rates and the entry probability specifies a jump Markov process.

Given this Markov model, the evolution of the average number of job sites of productivity c at time t, $f(c, t)$, obeys the following master equation:

$$\frac{\partial f(c,t)}{\partial t} = w_+(c-1)f(c-1,\ t) + w_-(c+1)f(c+1,\ t)$$

$$-w_+(c)f(c,\ t) - w_-(c)f(c,\ t) + p\delta_{c,1}, \tag{3.28}$$

Here, $\delta_{c,1}$ is 1 if $c = 1$ and 0 otherwise. This equation shows that the change in $f(c,\ t)$ over time is nothing but the *net* inflow to the state c.

---

6 Yukalov and Sornette (2012) in an interesting model similar to the one considered here shows that under certain conditions statistical outliers with abnormal size which they call "dragon-kings" can emerge as Bose-condensed droplets along with Zipf's law for the rest of members. The model can be easily reinterpreted as a model of the levels of productivity in the economy. Dragon-king corresponds to the abnormally high level of productivity. The study of such an outlier is left for further investigation.

We consider the steady state or the stationary solution of Eq. (3.28) such that $\partial f(c,\ t)/\partial t = 0$. The solution $f(c)$ can be readily obtained by using the boundary condition that $w_-(1)\ f(1) = p$:

$$f(c) = f(1)\prod_{k=1}^{c-1} \frac{w_+(c-k)}{w_-(c-k+1)}. \tag{3.29}$$

Here, we make an important assumption on the transition rates, w_+ and w_-. Namely, we assume that the probabilities of an increase and a decrease in productivity depend on the current level of productivity of the job site. Specifically, the higher the current level of productivity is, the larger the chance of unit productivity change. This assumption means that the transition rates can be written as $w_+(c) = a_+c^\alpha$ and $w_-(c) = a_-c^\alpha$, respectively. Here, a_+ and a_- are positive constants, and α is greater than 1. Under this assumption, the stationary solution (3.29) becomes

$$f(c) = \frac{f(1)}{1 - f(1)/C_{(\alpha)}} \frac{\left(1 - f(1)/C_{(\alpha)}\right)^c}{c^\alpha} \cong c^{-\alpha} e^{-c/c*}. \tag{3.30}$$

where

$$c^* \equiv \left(f(1)/C_{(\alpha)}\right)^{-1} \quad \text{and} \quad C_{(\alpha)} \equiv \sum_{c=1}^{\infty} c^\alpha f(c).$$

We use the relation $a_+/a_- = 1 - f(1)/C_{(\alpha)}$. The approximation in Eq. (3.30) follows from $n(1)/C_{(\alpha)} << 1$, and the exponential cutoff works as c approaches c^*. However, the value of c^* is practically large, and therefore, we observe the power law distribution $n(c) \propto c^{-\alpha}$ for a wide range of c despite the cutoff. Therefore, under the reasonable assumption, we obtain a power law distribution for job sites, f_j.

The above model can be understood easily with the help of an analogy with the formation of cities. Assume that $f(c,\ t)$ is the number of cities with population c at time t. $w_+(c)$ corresponds to a birth in a city with population c or an inflow into the city from another city. Similarly, $w_-(c)$ represents a death or an exit of a person moving to another city. These rates are the instantaneous probabilities that the population of a city with the current population c either increases or decreases by 1. They are, therefore, the entry and exit rates of *one person* times the population c of the city. In addition, a drifter forms his own one-person city with the instantaneous probability p. In this model, the dynamics of $f(c,\ t)$ or the average number of cities with population c, is given by Eq. (3.28). In the case of population dynamics, one might assume that the entry (or birth) and exit (or death) rates of a person, a_+ and a_-, are independent of the size of the population of the city in which the person lives. In that case, $w_+(c)$ and $w_-(c)$ become linear functions of c, namely, a_+c and a_-c. However, even in population dynamics, one might assume that the entry rate of a person into a large city is higher than its counterpart in a smaller one because of better job opportunities or the social attractiveness of such places as encapsulated in the words of the song, "bright lights, big city". The same may hold true for exit and death rates because of congestion or epidemics.

In turn, in the dynamics of productivity of job sites, both the "entry" and "exit" rates of an existing "productivity job site" are increasing functions of c, namely, the level of productivity in which that particular job site happens to be located; to be concrete, a_+c and a_-c. Thus, $w_+(c)$ becomes a_+c times c, which is equal to a_+c^2. Likewise, we obtain $w_-(c) = a_-c^2$. This is the case for the so-called Zipf law (Ijiri & Simon, 1975). Thus, under the reasonable assumption that the probability of a unit change in productivity is an increasing function of its current level c, we obtain a power law distribution for job site f_j.

Changes in productivity are caused not only by supply-side factors such as technical progress and the entry/exit of firms but also by sectoral shifts in demand. Indeed, Fay and Medoff (1985) documented such changes in a firm's labor productivity by way of changes in the rate of labor hoarding. Stochastic productivity changes as described in our Markov model certainly include technical progress, particularly in the case of an increase, but at the same time they embrace *allocative demand disturbances*. By their careful study of industry-level productivity and worker flows, Davis et al. (1996) found that while job creation is higher in industries with high total factor productivity growth, job destruction is not systematically related to productivity growth. Job *destruction* is actually highest in the industries in the top productivity growth quintile (Table 3.7, Davis et al., 1996, 52). This finding suggests the importance of *negative demand shocks* for job destruction.

Distribution of Productivity

Under reasonable assumptions, the distribution of potential job sites with high productivity follows a downward-sloping power law. However, the determination of employment by firms with various levels of productivity is another matter.

To fill potential job sites with workers is the firm's economic decision. The most important constraining factor is the level of demand facing the firm in the product market. Whatever the level, to fill potential job sites, the firm must either keep the existing workers on the job or make enough hiring efforts, including posting vacancy signs toward successful job matching. Such actions of the firms and job searches of workers are purposeful. However, micro shocks affecting firms and workers are unspecifiable. Then, how are workers actually employed at firms with various levels of productivity? This is the problem we considered in the previous section. In what follows, we will consider it in a more general framework.

The number of workers working at the firms with productivity,c_j, namely, n_j, is

$$n_j \in \{0, 1, \ldots, f_j\} \quad (j = 1, 2, \ldots K). \tag{3.31}$$

Here, f_j is the number of potential jobs with productivity c_j and puts a ceiling on n_j.[7] We assume that in the low productivity region, f_j is large enough, meaning that n_j is virtually unconstrained by f_j. In contrast, in the high productivity region, f_j constrains n_j, and its distribution is a power distribution, as we explained above.

Like many others, Postel-Vinay and Robin (2002) assume that there is no ceiling for job sites with high productivity. On this assumption as well as the assumption that the product market is perfectly competitive, they regard a decreasing number of workers with high productivity jobs as a consequence of less recruitment efforts made by high productivity firms than by low productivity firms. This is a strange interpretation. There is no reasonable reason why high-productivity firms make fewer recruitment efforts than low-productivity firms. A more plausible assumption is that firms are monopolistically competitive facing the downward-sloping individual demand curve rather than the price takers in the product market and that jobs with high productivity are limited in number by such demand constraints. The automobile industry is a high productivity industry. It would be unreasonable to argue that the level of employment in the industry is only the outcome of job matching and that a limited size of employment is due to a lack of the firms' recruitment efforts. Plainly, the size of car producers is given by their capacity, which is, in turn, determined by the level of demand for cars.

The total number of employed workers is simply the sum of n_j:

$$N = \sum_{j=1}^{K} n_j. \tag{3.32}$$

In the basic model, the total number of employed workers, N is exogenously given (Eq. (3.17)). In the extended model, N is assumed to be variable. N is smaller than the exogenously given total number of workers or labor force, L ($N < L$). The difference between L and N is the number of unemployed individuals, U:

$$U = L - N. \tag{3.33}$$

As in the basic model, firms are monopolistically competitive when facing the downward-sloping kinked individual demand curve. The firm's output is constrained by demand, which is conditioned by the level of aggregate demand, D (Eq. (3.22)). In the basic model, D is literally given. Accordingly, the total output is also given by Eq. (3.22) and is constant. In the present model, we more realistically assume that in accordance with fluctuations of aggregate demand, total output Y also fluctuates. Specifically, Y defined by

[7] When the number of potential jobs with high productivity is limited, behavior of economic agents necessarily becomes correlated; If good jobs are taken by some workers, it becomes more difficult for others to find such jobs. Garibaldi and Scalas (2010) suggest that we study the problem by Markov model with such constraints. Aoyama, Iyetomi, and Yoshikawa (2015) derives equilibrium distribution of workers across firms with different productivity under the constraint of limited number of jobs using the detail balance condition in the Ehrenfest-Brillouin model.

$$Y = \sum_{k=1}^{K} y_k = \sum_{k=1}^{K} c_k n_k \tag{3.34}$$

is now stochastic, and its expected value $< Y >$ is equal to the constant D. That is, we have

$$< Y >= D. \tag{3.35}$$

Aggregate demand constrains total output in the sense of its expected value.

Under this assumption, the probability of total output Y turns out to be exponential.[8] The density function $g(Y)$ is

$$g(Y) = \frac{e^{-\beta Y_i}}{\sum_i e^{-\beta Y_i}} \tag{3.36}$$

Obviously, Y constrained by aggregate demand D affects the distribution of workers, n_k (Eq. (3.34)). In the present model, the number of employed workers N is not constant but changes. Not only the level of aggregate demand D, the number of potential job sites with high productivity, f_j also constrains n_j. Under these assumptions, we seek the state that maximizes the probability P_n or Eq. (3.19).

Before we proceed, it is necessary to explain the *partition function* because it is rarely used in economics, but we will use it intensively in the subsequent analysis. When a stochastic variable Y is exponentially distributed, that is, its density function $g(Y)$ is given by Eq. (3.36), partition function Z is defined as

$$Z = \sum_i e^{-\beta Y_i}. \tag{3.37}$$

This function is extremely useful as a moment generating function. For example, we can easily find the first moment or the average of Y by differentiating $\log Z$ with respect to β.

[8] This result is obtained by the method of Gibbs' canonical ensemble. Gibbs established the statistical mechanics by introducing the concept of "canonical ensemble" which is a collection of macro states, Y in our present case. Suppose that there are K possible levels of Y denoted by $Y_1, \ldots Y_K$. For the moment, we reinterpret n_k as the number of cases where Y takes the value Y_k $(k = 1, \ldots, K)$. The sum of n_k, N is given. Then, n_k satisfies Eq. (3.17). We assume that the average of Y is equal to constant D: $\sum_{k=1}^{K} Y_k (n_k / N) = D$. Replacing c_k by Y_k/N, we observe that this equation is equivalent to Eq. (3.22). Thus, we can apply the exactly same entropy maximization as we did in the basic model in Sect. 3.3. It leads us to Eq. (3.36).

$$-\frac{d\log Z}{d\beta}=-\frac{d}{d\beta}\log\left(\sum_i e^{-\beta Y_i}\right)=-\frac{\sum_i(-Y_i)\,e^{-\beta Y_i}}{\sum_i e^{-\beta Y_i}}$$
$$=\sum_i Y\left(\frac{e^{-\beta Y_i}}{\sum_i e^{-\beta Y_i}}\right)=\sum_i Y_i g(Y_i)=E(Y_i) \tag{3.38}$$

As in the basic model, we want to find the state that maximizes the probability, P_n or Eq. (3.19). We have two macroconstraints, Eqs. (3.32) and (3.34). The total number of workers employed N is, however, not constant but variable. The aggregate output Y is also not constant but obeys the exponential distribution, namely, Eq. (3.36).

Because the level of total output depends on the total number of employed workers N, we denote Y_i as $Y_i(N)$. Then, the canonical partition function Z_N can be written as

$$Z_N=\sum_i e^{-\beta Y_i(N)}. \tag{3.39}$$

Using Eq. (3.34), we can rewrite this partition function as follows:

$$Z_N=\sum_{\{n_i\}}\exp\left(-\beta\sum_{i=1}^{K}c_i n_i\right). \tag{3.40}$$

It is generally difficult to carry out the summation with respect to $\{n_i\}$ under constraint (3.32), namely, $N=\sum n_i$. Rather than taking N as given, we better allow N to be variable as we do here and consider the following extended partition function Φ. It is called the grand canonical partition function in physics. With the Lagrangean multiplier for N, μ, this partition function is defined as

$$\Phi=\sum_{N=0}^{\infty}z^N Z_N \tag{3.41}$$

where

$$z=e^{\beta\mu}. \tag{3.42}$$

As the Lagrangean multiplier for N, the parameter μ is the marginal contribution of an addition employment to the entropy of the macroeconomy.[9] Because entropy corresponds one to one to the level of aggregate demand, as shown in Fig. 3.5, it measures the marginal product of a worker who newly acquired a job out of the pool

[9] μ is called the chemical potential in physics, and measures the marginal contribution in terms of energy of an additional particle to the system under investigation.

of unemployment. Thus, μ plays a role similar to the reservation wage in standard models. When μ is high, the unemployed worker is "choosy", and vice versa. In this sense, μ is a barrier between unemployment and employment. However, there is a fundamental difference between the reservation wage in standard theory and μ. In standard models, the reservation wage R is absolute in the sense that no worker is willing to work with the wages below R. This concept is useful only if R can be well defined and known. As we pointed out in Sect. 3.2, equilibrium search theory effectively presumes the representative worker so that the reservation wage R can be uniquely defined.

In contrast, μ in the present analysis plays only a relative role as a barrier between unemployment and employment. Each worker has his/her own "reservation job attractiveness", which we do not observe. μ represents the *average* reservation job attractiveness of a heterogeneous group of unemployed workers. We will explain the meaning of μ later. For the moment, it is important to note that depending on the level of aggregate demand, some workers accept jobs with productivity below μ.

Substituting Eq. (3.40) into Eq. (3.41), the grand canonical partition function Φ becomes as follows:

$$\Phi = \sum_{N=0}^{\infty} z^N \sum_{nj} \exp\left\{ -\beta \sum_j n_j c_j \right\} \qquad \text{where } z = e^{\beta\mu} \tag{3.43}$$

Using the definitions of z, (3.42), and N, (3.32), we have

$$\Phi = \sum_{N=0}^{\infty} e^{\beta\mu(n_1+\ldots+n_K)} \sum_{n_j} \exp\{-\beta \sum_j n_j c_j\} = \prod_{j=1}^{K} \sum_{nj} \exp[\beta(\mu - c_j)n_j]. \tag{3.44}$$

Because there is a ceiling f_j for n_j (constraint (3.31)), (3.44) can be rewritten as follows:

$$\Phi = \prod_{j=1}^{K} [1 + e^{\beta(\mu-c_j)} + \cdots e^{f_j\beta(\mu-c_j)}] \tag{3.45}$$

With this grand canonical partition function Φ, we can easily obtain the expected value of the total number of employed workers N, $< N >$ by differentiating $log\Phi$ with respect to μ, which corresponds to the reservation wage of the unemployed worker. This can be seen by differentiating (3.41) and noting the definition of z, (3.42).

$$\begin{aligned}\frac{1}{\beta}\left[\frac{\partial}{\partial\mu}\log\,\Phi\right] &= \frac{1}{\beta}\left[\frac{\partial}{\partial\mu}\log\left(\sum_{N=0}^{\infty} e^{\beta\mu N}\, Z_N\right)\right] \\ &= \frac{1}{\beta}\left[\frac{\beta\sum_{N=0}^{\infty} e^{\beta\mu N}\, Z_N}{\sum_{N=0}^{\infty} e^{\beta\mu N}\, Z_N}\right] = < N > .\end{aligned} \tag{3.46}$$

In the present case, Φ is actually given by Eq. (3.45). Therefore, we can find $< N >$ as follows.

$$\begin{aligned}\langle N \rangle &= \frac{1}{\beta}\left[\frac{\partial}{\partial \mu}\log \Phi\right] \\ &= \frac{1}{\beta}\sum_{j=1}^{K}\frac{\partial}{\partial \mu}\left\{\log\left(1 + e^{\beta(\mu-c_j)} + \cdots + e^{(f_i\beta(\mu-c_j))}\right)\right\} \\ &= \sum_{j=1}^{K}\left[\frac{e^{-(f_j-1)\beta(\mu-c_j)} + 2e^{-(f_j-2)\beta(\mu-c_j)} + \cdots + f_j}{e^{-f_j\beta(\mu-c_j)} + e^{-(f_j-1)\beta(\mu-c_j)} + \cdots + 1}\right]\end{aligned} \tag{3.47}$$

The expected value of the number of workers employed on the job sites with productivity c_j, $< n_j >$ is simply the corresponding term in the summation of $< N >$ or Eq. (3.47).

$$\langle n_j \rangle = \left[\frac{e^{-(f_j-1)\beta(\mu-c_j)} + 2e^{-(f_j-2)\beta(\mu-c_j)} + \cdots + f_j}{e^{-f_j\beta(\mu-c_j)} + e^{-(f_j-1)\beta(\mu-c_j)} + \cdots + 1}\right] \tag{3.48}$$

Equation (3.48) determines the distribution of workers across job sites with different levels of productivity in stochastic macroequilibrium.

Figure 3.8 shows how this model works. On the one hand, there are dynamics of the creation and destruction of potential job sites with various levels of productivity (Fig. 3.8a). We presented a Markov model that leads us to the power-law tail of productivity distribution in the steady state. At the same time, there is another dynamics of job matching (Fig. 3.8b). The evolution of two dynamics determines the level of unemployment and the distribution of workers at job sites with various productivities, Eq. (3.48).

First, we consider the distribution of productivity. Most importantly, the distribution is fundamentally conditioned by aggregate demand. When the level of aggregate demand is high, it is more likely that high productivity firms employ more workers. They keep the existing workforce and attract not only the unemployed but also workers currently on inferior jobs. As Okun (1973) vividly illustrates, "high pressure economy provides people with a chance to climb ladders to better jobs." People actually climb ladders in such circumstances.

Note that in this model, most workers climb ladders to better jobs or flow toward higher productivity jobs without experiencing unemployment. Haltiwanger et al. (2008) indeed demonstrate that workers tend to flow, on net, from low-wage to high-wage firms via direct job-to-job moves and that this movement is strongly procyclical. Although the present model focuses on productivity rather than wages, it is known that high-paying firms are also more productive firms (Abord, Kramarts, & Margolis, 1999; Abord et al., 2005). In passing, productive firms are not necessarily large firms. That is what Haltiwanger et al. (2008) found for the U.S. economy. As they point it out, small firms consist of two groups, young and mature, and some small, young firms are highly productive. The present model considers the distribution of productivity without referring to firm size.

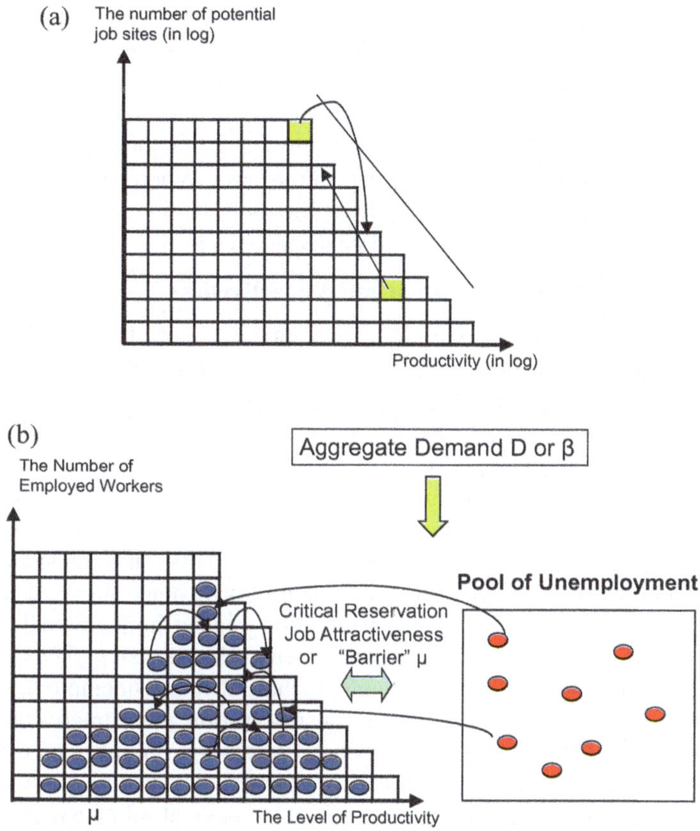

Fig. 3.8 Model of Stochastic Macroequilibrium

(a) Dynamics of Creation and Destruction of Potential Jobs
(b) Stochastic Macroequilibrium
Note Both productivity and the number of potential job sites are in the natural logarithm. The straight line as drawn in (a) means that the distribution of productivity is power-law.

Source Yoshikawa (2015)

It is useful to consider $< n_j >$ in (3.48) under extremely high and low levels of aggregate demand, respectively. This exercise clarifies the meaning of a "barrier" between unemployment and employment μ.

The first is the case of the extremely high aggregate demand. In this case, β is $-\infty$ (Table 3.1). Given (3.48), for $c_j > \mu$, $< n_j >$ becomes f_j, whereas for $c_j < \mu$, $< n_j >$ is zero. That is, when aggregate demand is extremely high, no worker takes jobs whose productivity is lower than μ, whereas employment at job sites with productivity greater than μ is constrained only by capacity, f_j. Note that only when the aggregate demand is extremely high does μ become equivalent to the standard reservation wage. When $c = \mu$, $< n_j >$ is equal to $f_j/2$.

In contrast, when aggregate demand is extremely low ($\beta =-0$), $< n_j >$ becomes equal to $f_j/2$ irrespective of c_j and μ. Note that $< n_j >$ is not zero for c_j below μ. In fact, because f_j is very large for low productivity jobs, a majority of workers obtain jobs whose productivity c_j is lower than μ in this case. When aggregate demand is low, a number of workers take "odd jobs". The more realistic case with the finite absolute value of β ($-\infty < \beta < 0$) will be discussed shortly.

Unemployment and the Reservation Job Attractiveness

The distribution of employed workers across job sites with different productivity, Eq. (3.48), simultaneously determines the level of unemployment because the number of unemployed U is simply the difference between the exogenously given labor force, L, and the number of employed workers, N, which is nothing but the sum of $< n_j >$. Recall that the expected value of N, $< N >$ is given by (3.47).

When aggregate demand D rises, more workers work at job sites with higher productivity, while at the same time, unemployment decreases. On the other hand, an increase in μ *ceteris paribus* raises unemployment. This can be most clearly seen in the case of the extremely high aggregate demand. As we noted above, in such a case ($\beta = -\infty$), μ plays a role similar to the reservation wage in the standard theory in the sense that no worker takes a job whose productivity is lower than μ. Thus, it is trivial to see that an increase in μ lowers employment and raises unemployment. More generally, however, we must recognize that reservation job attractiveness differs across workers and that it cannot be reduced to a single number of a particular variable, such as wages. μ in this model is a "barrier" between unemployment and employment, but unlike the reservation wage, it is not an absolute barrier. Given μ, when aggregate demand is low, workers actually work at job sites whose productivity is lower than μ. As we will see shortly, the effect of a change in μ on unemployment itself depends on the level of aggregate demand.

Unemployment, which arises because of frictions and uncertainty in the labor market, is by definition a job search. Each worker has his/her own reservation job attractiveness. However, frictions and uncertainty depend on the level of aggregate demand. When aggregate demand is high, high productivity firms make many job openings so that workers can find attractive jobs easily. When aggregate demand is low, just the opposite holds true. In this sense, some of the so-called "frictional" unemployment is actually caused by aggregate demand deficiency.

Thus, *the definition of frictional unemployment, let alone natural unemployment* (Friedman, 1968), *must be necessarily ambiguous unless the level of aggregate demand is unrealistically high.*

A Numerical Example

With the help of a simple numerical example, we can better understand what Eq. (3.48) looks like and how unemployment is determined in the present model. In this example, the level of productivity is assumed to be $c_1 = 1, \ldots c_{200} = 200$. μ is set equal to 25. The labor force L is assumed to be 630. The number of potential jobs or the ceiling at each productivity level, f_j, is assumed to be constant at 10 for $c_1, \ldots c_{50}$, while it declines for c_j ($j = 50, \ldots, 200$) as c_j increases. Specifically, for c_j ($j = 50, \ldots, 200$), f_j obeys a power distribution: $f_j \sim 1/c_j^2$. This assumption means that low productivity jobs are potentially abundant, whereas high productivity jobs are limited. In the following figures, the number of potential jobs or "capacity" is shown by a dotted line.

Figure 3.9 shows labor productivity distributions (Eq. (3.48)) under (A) extremely high and (B) low levels of aggregate demand. As explained previously, when aggregate demand is extremely high ($\beta = -\infty$), high-quality jobs are fully occupied at the ceiling level, while no worker takes jobs whose productivity is lower than $\mu = 25$. In contrast, when aggregate demand is extremely low ($\beta = 0$), employment is equal to $f_j/2$ regardless of the level of productivity c_j. A number of workers take jobs with productivity lower than μ, while at the same time, high-quality job sites are left unfilled.

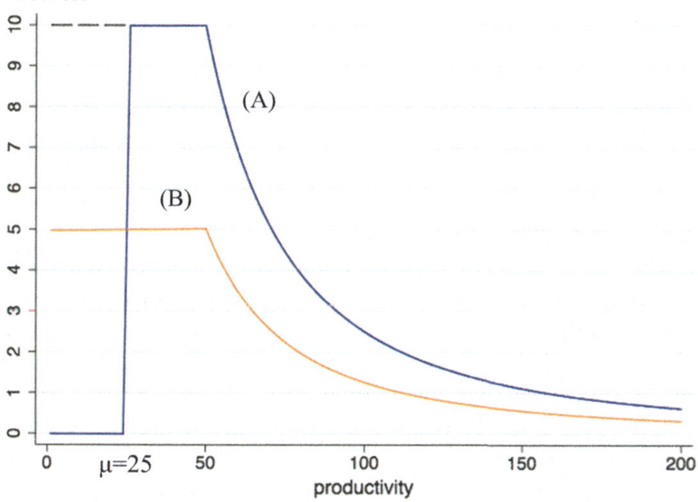

Fig. 3.9 Job quality distribution under the extremely high and low levels of aggregate demand

Note (A) Extremely high aggregate demand ($\beta \cong -\infty$),
(B) Extremely high Aggregate Demand ($\beta \cong 0$).

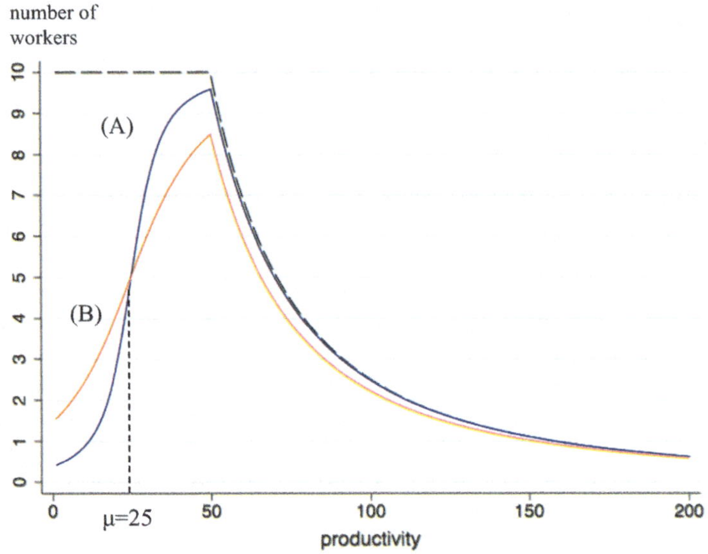

Fig. 3.10 Distribution of productivity and aggregate demand

Note (A) High aggregate demand ($\beta = -0.05$),
(B) Low Aggregate Demand ($\beta = -0.02$). See the main text for details.

Figure 3.10 shows two realistic cases; Case A corresponds to high aggregate demand, whereas Case B corresponds to low aggregate demand. Specifically, β is assumed to be (A) -0.05 and (B) -0.02. As explained in Table 3.1, Case (A) $\beta = -0.05$ corresponds to high demand, whereas Case (B) $\beta = -0.02$ corresponds to low demand. In Fig. 3.10, we observe that n_j increases up to $j = 50$ and then declines from $j = 50$ –200 in both cases.

Whenever possible, workers strive to obtain better jobs offered by firms with higher productivity. That is why the number of workers n_j increases as the level of productivity rises in the relatively low productivity region. Note that the number of potential jobs or the ceiling is constant (virtually, no ceiling) in this region, and yet that n_j increases as the level of productivity c_j rises. The number of workers n_j turns out to be a decreasing function of productivity c_j in the high productivity region simply because the number of potentially available jobs f_j declines as c_j rises. Note, however, that n_j is not equal to f_j; n_j is strictly smaller than f_j. The single-peaked distributions shown in Fig. 3.10 are broadly consistent with the observed pattern of productivity dispersion (Fig. 3.7).

When aggregate demand D increases, the distribution shifts in such a way that more workers work at high productivity job sites and fewer workers work at low productivity job sites. Figure 3.11 shows how employment at job sites with different productivity depends on the level of aggregate demand. When aggregate demand

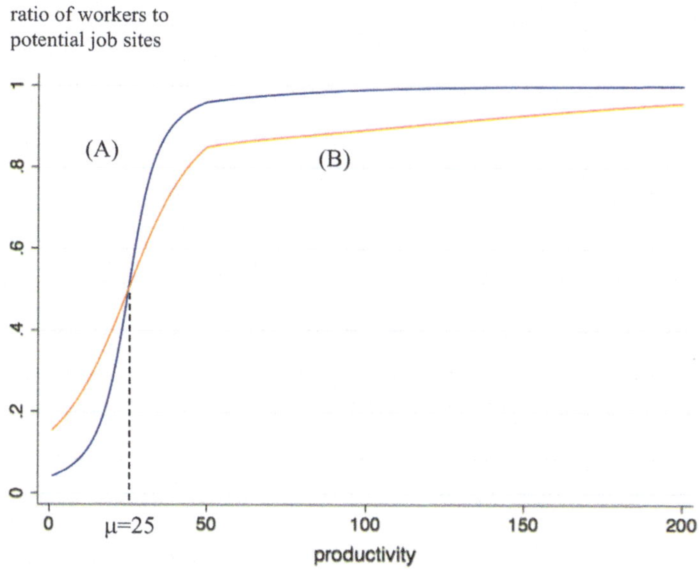

Fig. 3.11 Percentage of potential job sites occupied by employed workers.

Note (A) High aggregate demand ($\beta = -0.05$),
(B) Low Aggregate Demand ($\beta = -0.02$). See the main text for details.

is high, the percentage of filled jobs with high productivity is high, whereas the percentage of filled jobs with low productivity (to be precise, lower than μ) is low, and vice versa.

When aggregate demand increases, the number of employed workers N, which corresponds to the area below the distribution curve, increases. Specifically, N is 618 in case (A) while it is 582 in case (B). This means that given labor force $L = 630$, the unemployment rate $U/L = (L - N)/L$ declines when aggregate demand D rises. In this example, the unemployment rate is 1.8% in case (A), while it is 7.6% in case (B).

Table 3.2 shows (a) the unemployment rate, (b) the share of workers on job sites with productivity higher than μ, and (c) the corresponding share of workers on job sites with productivity lower than μ for various levels of aggregate demand or β. When aggregate demand rises, the unemployment rate declines, while at the same time, the share of high-quality jobs increases.

Using this numerical example, we can also consider the effects of a change in reservation job attractiveness μ on employment/unemployment. In the base model, μ is set equal to 25. Table 3.3 shows how employment and unemployment change when μ increases from 25 to 28. μ corresponds to the reservation wage in standard models so that an increase in μ raises the unemployment rate. However, Table 3.3 shows that the extent to which the unemployment rate rises depends crucially on the level of aggregate demand. The higher the level of aggregate demand is, the more

Table 3.2 Aggregate demand, job quality, and unemployment

Aggregate demand	← Low			High →		
β	−0.001	−0.01	−0.02	−0.03	−0.05	−0.1
Unemployment rate (U/L) %	29.2	15.9	7.6	4.1	1.8	0.9
Share of Employment with probability $c_j > \mu$ (%)	74	83	88	91	94	97
Share of employment with probability $c_j < \mu$ (%)	26	17	12	9	6	3

Note μ = The critical "reservation job attractiveness." See the main text for details

Table 3.3 The effects of a change in the reservation job attractiveness μ on unemployment depends on the level of aggregate demand

Aggregate demand	← Low			High →	
β	−0.001	−0.005	−0.01	−0.02	−0.05
(1) Employment when μ = 25	446	487	530	582	618
(2) Employment when μ = 28	444	477	513	557	590
(3) Change in Employment whenμ increased from 25 to 28 ((2)–(1))	−2	−10	−17	−25	−28
(4) Change in unemployment rate ((3)/L = 630)	0.3	1.6	2.7	4.0	4.5

significant the effect of an increase in μ on the unemployment rate. It is important to recognize that the effects of a change in μ on unemployment decrease when the level of aggregate demand is low. In other words, in deep recessions, the "reservation wage" does not play a major role as a determinant of unemployment.

3.6 The Principle of Effective Demand

Keynes (1936) argued that the aggregate demand determines the level of output in the economy as a whole. Factor endowment and technology may set a ceiling on aggregate output, but the actual level of output is effectively determined by aggregate demand. Factor endowment and technology change only slowly, but aggregate demand can change swiftly. That is why the level of aggregate demand rather than technology becomes the effective constraint for the macroeconomy. Unemployment is obviously a problem of the labor market, but Keynes (1936) argued that it is basically caused by demand deficiency in the market for goods and services. Changes in aggregate demand determine cyclical changes in unemployment.

Keynes's theory has long been debated in terms of unemployment or "involuntary" unemployment. Although unemployment is one of the most important economic problems in any society, focusing only on unemployment is inadequate for the purpose of providing microfoundations for Keynesian economics. The real issue is whether there is any room for mobilizing labor to high productivity jobs, firms, or sectors. The famous Okun's law demonstrates that there is always such room in the economy (Okun, 1963); see Syverson (2011) for more recent research on productivity dispersion. Unemployment is actually only a partial measure of the underutilization of labor. We must focus on the distribution of labor productivity. It is fundamentally conditioned by aggregate demand.

The utilization rate of the production factor depends on demand. John Stuart Mill noticed this over a hundred and fifty years ago:

> When we have thus seen accurately what really constitutes capital, it becomes obvious, that of the capital of a country, there is at all times a very large proportion lying idle. The annual produce of a country is never any thing approaching in magnitude to what it might be if all the resources devoted to reproduction, if all the capital, in short, of the country, were in full employment.
>
> If every commodity on an average remained unsold for a length of time equal to that required for its production, it is obvious that, at any one time, no more than half the productive capital of the country would be really performing the functions of capital …
>
> From the considerations which we have now adduced, it is obvious what is meant by such phrases as a brisk demand, and a rapid circulation. There is a brisk demand and a rapid circulation, when goods, generally speaking, are sold as fast as they can be produced. There is slackness, on the contrary, and stagnation, when goods, which have been produced, remain for a long time unsold. In the former case, the capital which has been locked up in production is disengaged as soon as the production is completed; and can be immediately employed in further production. In the latter case, a large portion of the productive capital of the country is lying in temporary inactivity.
>
> From what has been already said, it is obvious, that periods of "brisk demand" are also the periods of greatest production: the national capital is never called into full employment but at those periods (Mill, 1844).[10]

Our economy truly experiences occasional *aggregate demand shocks*. A case in point is Japan's post-Lehman "Great Recession" during 2008–09. Globally, it was triggered by a financial crisis. Japan's financial system during the period was sound and safe,

[10] In spite of this statement, Mill (1844) did not regard the state of "full employment" desirable, for the following reason. "This, however, is no reason for desiring such times; it is not desirable that the whole capital of the country should be in full employment. For the calculations of producers and traders being of necessity imperfect, there are always some commodities which are more or less in excess, as there are always some which are in deficiency. If therefore the whole truth were known, there would always be some classes of producers contracting, not extending, their operations. If all are endeavoring to extending them, it is a certain proof that some general delusion is afloat. The commonest cause of such delusion is some general, or very extensive, rise of prices (whether caused by speculation or by the currency) which persuades all dealers that they are growing rich" (p. 67). Here, Mill sounds like Hayek.

however. Japan's Great Recession was basically caused by an unprecedented fall in exports. A sudden fall of exports was an exogenous real demand shock to the Japanese economy. It is absurd to call it a "technology" or "productivity" shock in the ordinary sense of words. The principle of effective demand is alive and well! We will formally show this in the next chapter.

The challenge to Keynesian economics is to provide a microfoundation for the principle of effective demand. The natural micro picture underneath Keynesian economics is monopolistic competition of firms facing the downward sloping individual demand curve, certainly not perfect competition in the product market (Solow, 1986). Negishi's (1979) model of the general equilibrium of monopolistically competitive firms with *the kinked individual demand curve* provides a neat microfoundation for what Tobin (1993) called the Old Keynesian view according to which "quantities determine quantities." This model, however, abstracts itself from the frictions and uncertainty present in the labor market.

Standard equilibrium search theory has filled a gap by explicitly considering frictions and matching costs in the labor market. However, in the name of productivity shocks, it tacitly presumes that the product market is perfectly competitive in the sense that the firm's individual demand curve is flat. It is simply wrong for the purpose of studying *cyclical* changes in the utilization rate of labor.[11] It is indeed curious that standard search theory places so much emphasis on frictions and uncertainty in the labor market, while at the same time, it is content with the naïve assumption of perfect competition in the goods market.

In the real economy, firms face a downward sloping individual demand curve with a kink (Negishi, 1979). This theoretical framework is neat. However, the equilibrium is indeterminate in the sense that we never know how the aggregate demand is distributed across firms or job sites with different levels of productivity. We are unable to know the micro behaviors of workers and firms, either.

A model of stochastic macroequilibrium based on the principle of statistical physics provides a solution. The concept of stochastic macroequilibrium is motivated by the presence of all kinds of unspecifiable micro shocks. At first, one might think that allowing all kinds of unspecifiable micro shocks leaves so many degrees of freedom that almost anything can happen. However, the methods of statistical physics—the maximization of entropy under macroconstraints—actually provide us with the quantitative prediction about the equilibrium distribution of productivity, namely, Eq. (3.48).

It is extremely important to recognize that the present approach does not regard the behaviors of workers and firms as random. They certainly maximize their objective

[11] Under the assumption of perfect competition, cyclical driving forces are identified with changes in the average productivity (See, for example, Shimer, 2005, 2010, and Hagedorn and Manovskii, 2008). This exercise is essentially the real business cycle (RBC) theory (Kydland & Prescott, 1982). Shimer (2005)'s comparative static analysis however, in effect, demonstrates that a change in labor productivity a la RBC cannot reasonably explain the empirically observed magnitude of fluctuations of the unemployment and vacancy rates. There is, in fact, a strong case that cyclical changes are caused by demand shocks rather than productivity shocks (Mankiw, 1989; Summers, 1986; Iyetomi et al., 2011). We will demonstrate it in Chap. 4.

functions perhaps dynamically in their respective stochastic environments. Of course, as behavioral economics has demonstrated, real optimization may be different from utility/profit maximization, as traditional economics takes as an axiom. However, that is not a crucial point any way for the purpose of macroeconomics.

The maximization of entropy under the aggregate demand constraint (3.23) balances two forces. On the one hand, whenever possible, workers are assumed to move to better jobs that are identified with job sites with higher productivity. We never know which jobs are how attractive to which workers. Firms make efforts to hire good workers under demand constraints in the goods market. Again, we never know which workers are how good to which firms. It is the outcome of successful job matching resulting from the worker's search and the firm's recruitment. When the level of aggregate demand is high, this force dominates because the demand for labor of high productivity firms is high.

However, as the aggregate demand decreases, the number of possible allocations consistent with the level of aggregate demand increases. More workers are forced to be satisfied with or look for low-productivity jobs. Randomness, which plays a crucial role in our analysis, basically comes from the fact that demand constraints in the product market facing firms with different productivity and optimizing behaviors of workers and firms under such constraints are so complex and unspecifiable that those of us who analyze the macroeconomy must take micro behaviors as random. The method is straightforward and does not require any arbitrary assumptions on the behavior of economic agents.

When the level of aggregate demand is high, it is most likely that high productivity firms keep more workers on the job[12] and make more aggressive hiring efforts than in the period of low demand. Workers are certainly aware of such a change. It is demonstrated by the fact that quit rates are higher in high-demand periods although the employed workers are treated better in such periods. Whether better jobs are truly offered and workers move to those jobs ultimately depends on the level of aggregate demand. Our analysis demonstrates that the most likely outcome of random matching of firms and workers is given by Eq. (3.48), which depends on aggregate demand. It broadly coincides with the empirically observed distribution of productivity.

We emphasize that *frictions and uncertainty in the labor market are not exogenously given but depend crucially on the aggregate demand. There is no such thing as the natural unemployment defined by Friedman (1968) because the aggregate demand fundamentally conditions the rate of successful matching*. We must also note that the role of reservation job attractiveness, μ, in the model also depends crucially on aggregate demand. It is shown that in a deep recession, reservation job attractiveness (or wages) plays only a relatively small role as a determinant of unemployment.

---

[12] To be precise, more workers are on the "job sites" with higher productivity because productivity may differ across job sites within a single firm. Strictly speaking, the method of statistical physics should apply to "job sites" rather than firms. A firm is a cluster of job sites with different productivity levels.

To repeat, the level of aggregate demand is the ultimate factor conditioning the outcome of random matching of workers and monopolistically competitive firms. By so doing, it changes not only unemployment but also the distribution of productivity and, as a consequence, the level of aggregate output. This is the market mechanism beneath Keynes's principle of effective demand.

Evidence

The distribution of productivity actually changes as the level of aggregate demand changes. The postLehman "Great Recession" provides us with an excellent opportunity to explore the effects of negative aggregate demand shock. The unemployment rate certainly rose. The relatively low and cyclically insensitive Japan's unemployment rate was 3.6% as of July 2007, but after the global financial crisis, it rose to 5.5% by July 2009. During the same period, the unemployment rate rose from 4.6 to 9.8% in the US. Unemployment is not the whole story, however. The distribution of productivity also changes.

Figure 3.12a–c compare the distributions of productivity *before* and *after* the Lehman crisis, namely, 2007 and 2009; (a) total, (b) manufacturing sector, and (c) nonmanufacturing sector. As our theory indicates, the distribution as a whole, in fact, tilts toward lower productivity in a severe recession. Figure 3.12 shows that the tilt of the distribution toward low productivity is more conspicuous for the manufacturing industry than for the nonmanufacturing industry. This is because in Japan, the 2009 recession after the sudden bankruptcy of the Lehman Brothers was basically caused by a fall of exports, and exports consist mainly of manufactured products such as cars. We can observe, however, that the distribution tilts toward low productivity for the nonmanufacturing industry as well, particularly in the high productivity region. This is, of course, because a fall of demand in the manufacturing sector spills over to the nonmanufacturing sector. Figure 3.12 demonstrates that the distribution of productivity indeed changes in the way as suggested in the theory of stochastic macroequilibrium.

Macrodisequilibrium

The concept of stochastic macroequilibrium assumes that the macroeconomy is in equilibrium in the sense of statistical physics. There is, however, some evidence suggesting that the macroeconomy may not be in equilibrium but consist of subsystems. Aoyama et al. (2015), for example, using the Japanese data described earlier, shows that the estimated β for manufacturing and nonmanufacturing sectors are different (Table 3.4). The nonmanufacturing sector is subject to lower aggregate demand than the manufacturing sector.

This finding does not mean that the theory of stochastic macroequilibrium is invalid but does suggest that the macroeconomy may consist of several sectors facing different levels of aggregate demand within each sector. The problem awaits further investigation.

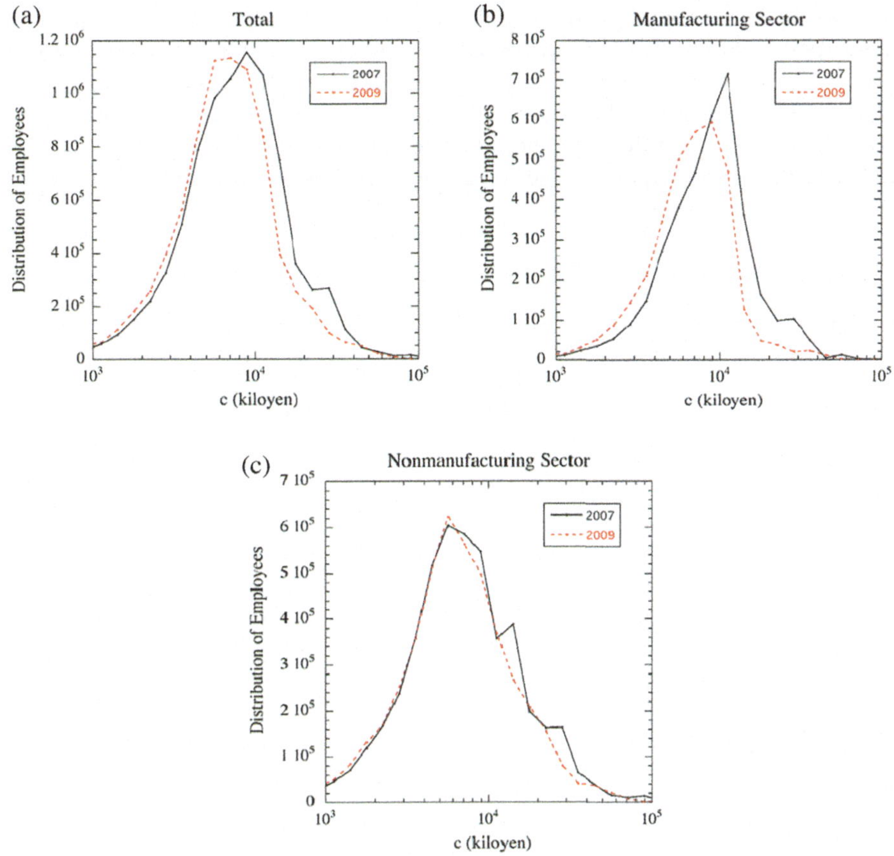

Fig. 3.12 Distribution of productivity
(a) Total
(b) Manufacturing sector
(c) Nonmanufacturing sector

Table 3.4 Estimated β (10^{-3})

	All	Manufacturing	Nonmanufacturing
β	−4.32	−8.68	−2.13

Source Aoyama et al. (2015)

Conclusion

Contrary to many economists' beliefs, the old principle of effective demand has a solid microfoundation. The market mechanism beneath Keynes's principle of effective demand is the general equilibrium of monopolistic competition coupled with search by workers and firms under friction and uncertainty. It is a cliché that the Keynesian problem of unemployment arises because prices and wages are inflexible. Tobin (1993), in fact, Keynes (1936) himself argued that the principle of effective demand holds true regardless of flexibility of prices and wages. This is because real demand changes much more swiftly than technology, and adjustment in terms of quantity is faster than that by way of changes in prices. Keynesian economics, in effect, claims that in the short run, aggregate demand conditions the matching of workers and firms and thereby determines the utilization of labor and the level of output in the macroeconomy. The logic does not depend on details of economic agents' micro behavior.

Chapter 4
Business Cycle: The Role of Aggregate Demand

4.1 Introduction

What causes business cycles? A glance at Haberler (1964) reveals that all kinds of theories had already been advanced by the end of the 1950s. In recent years, macroeconomists have been so confident of their skillful policy management as to hail the "Great Moderation". Lucas (2003), in his Presidential Address to the meeting of the American Economic Association, made the following declaration:

> Macroeconomics was born as a distinct field in the 1940's, as a part of the intellectual response to the Great Depression. The term then referred to the body of knowledge and expertise that we hoped would prevent the recurrence of that economic disaster. My thesis in this lecture is that macroeconomics in this original sense has succeeded: Its central problem of depression prevention has been solved, for all practical purposes, and has in fact been solved for many decades (Lucas, 2003, 11).

Ironically, when Lucas delivered this address, the American economy had been on the steady road to the worst postwar recession, perhaps even depression. Business cycles are still with us and await further investigation.

The real business cycle (RBC) theory pioneered by Kydland and Prescott (1982), arguably the most dominant school today, regards changes in total factor productivity (TFP) as the primary cause of business cycles. This theory is basically the neoclassical equilibrium theory and minimizes the role of the government's stabilization policy. Keynesian theory, in contrast, attributes business cycles to fluctuations in real aggregate demand (Summers, 1986; Mankiw, 1989; Tobin, 1993) and assigns the government and the central bank the substantial role of stabilization. The fundamental problem, namely, what causes business cycles, remains unresolved.

As Fig. 4.1 shows, measured TFP is highly correlated with the ups and downs of the macroeconomy. The problem is causality. RBC takes it that changes in TFP cause fluctuations in the economy. On the other hand, the Keynesian argues that a sizable portion of measured TFP is merely a reflection of economic fluctuations. Namely, the causality runs from business cycles to measured TFP.

H. Yoshikawa, *Reconstruction of Macroeconomics: Methods of Statistical Physics, and Keynes' Principle of Effective Demand*, Advances in Japanese Business and Economics 3, https://doi.org/10.1007/978-981-19-5264-7_4

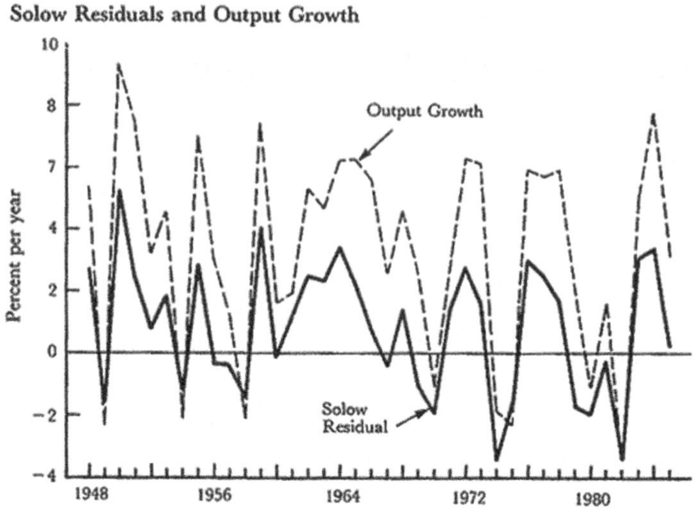

Fig. 4.1 TFP and economic growth rate. *Source* Mankiw (1989)

Suppose that we measure the labor productivity of someone who sorts out cars in a parking lot. The reasonable measure would be the number of cars he sorts out per hour. The measured labor productivity fluctuates, but it is nothing but the number of cars in and out having nothing to do with technical change, innovation or whatever else. When only a small number of cars are present, the worker becomes idle, and his work intensity decreases. The problem is that changes in work intensity are hard to measure. The unmeasured fluctuations of work intensity bring about a high correction of measured productivity and economic activity (the number of cars in this example). The same holds true for TFP. The negative growth of TFP often observed in Fig. 4.1 is "technical regress," but what is it truly?

The formal approach to study this problem is to explore the relative importance of technology shocks on the one hand and demand shocks on the other as the sources of business cycles by making various identifying restrictions on macro time-series models; see Shapiro and Watson (1988), Blanchard and Quah (1989), Francis and Ramey (2005) and the works cited therein. Although the variables used and methods differ in detail, they all share the same key identifying restriction. Namely, the level of output (or real GDP) is determined in the long run only by supply shocks such as technology shocks. Conversely, the long-run effect of demand shocks on output is assumed to be nil. On this assumption, nondirectly observable disturbances are sorted into technology shocks and demand shocks.[1]

[1] Shapiro and Watson (1988) consider labor supply shocks as well as technology shocks, and obtain the "surprising" result that the labor supply shocks account for at least 40% of output variation at all horizons. However, Hall (1988) shows that this finding is neither surprising nor plausible. Specifically, as Hall criticizes, Shapiro and Watson (1988) make the implausible identifying assumption that all the contemporaneous comovements of output and work hours be attributed to shifts in labor

A majority of works routinely make this assumption without questioning its validity. However, we maintain that the assumption is not actually so tenable as is commonly believed. The problem is that in the study of business cycles, technology shocks are usually taken exogenously as typified by real business cycle theory (Kydland & Prescott, 1982). Here, we must observe that modern microfounded macroeconomics is curiously inconsistent in that while new growth theory places so much emphasis on the endogeneity of technical progress, business cycle theory is naively content with taking advances in technology as exogenous. New technology or technical progress is certainly not a free lunch! First, it must be generated by R&D investment. Once new technology is identified, it can augment production only by way of investment, which embodies such new technology. Now, investment, including R&D investment, is, of course, much affected by cyclical fluctuations. Monetary factors as well as real factors affect investment. Thus, to the extent that new technology is embodied in investment, demand shocks are bound to affect the pace of technical progress. Therefore, we cannot naively accept the standard identifying restriction that demand shocks do not leave the permanent effects on the level of output.

One way to proceed without this questionable assumption is to take advantage of disaggregated data. The representative works on business cycles cited earlier all use the macro variables. However, business cycles, as conspicuous macro phenomena, are characterized by output movements across broadly defined sectors and types of goods moving together (Mitchell, 1951). In Mitchell's terminology, they exhibit high conformity. It is then most natural to seek possibly unobserved common factors underlining sectoral comovements as the basic factors causing business cycles. Moreover, the separate effects of such unobserved factors on production, shipments and inventory are expected to provide us with valuable information on the causes of business cycles.

Factor models precisely fit our purpose (Sargent & Sims, 1977; Stock & Watson, 1998, 2002). Stock and Watson (1998, 2002), based on this method, developed the diffusion index to be used for improving forecasting of cyclical fluctuations. In the next section, we take the same approach for a different purpose. By examining the relationship between the production, shipments, and inventory of 21 types of goods, we explore what causes business cycles. We also show that the model estimated based on the pre-Lehman crisis (September 15th, 2008) can reasonably explain the postLehman (2008–2009) recession of the Japanese economy. Our findings are in favor of the old Keynesian view (Tobin, 1993).

---

supply, and rule out the alternative assumption, perhaps more plausible to many, that the moving force operates from output to hours in the short-run. This paper indeed demonstrates that output is basically determined by real demand in business cycles.

4.2 Analysis Based on Industry Data

We use the monthly data of the Indices of Industrial Production (IIP) compiled by the Ministry of Economy, Trade and Industry (METI) of Japan. They are indices of industrial production, producer shipments, and producer inventory of finished goods. Details of IIP are on the website of METI (2008). Here, we simply note that the indices are compiled based on nearly 500 items (496 for production and shipment indices; 358 for inventory). We use the data for the 240 months from January 1988 to December 2007.

We denote the IIP by $S_{\alpha,g}(t_j)$, where $\alpha = 1, 2$, and 3 denote production (value added), shipments, and inventory, respectively. Additionally, $g = 1, 2..., 21$ denotes 21 categories of goods, $t_j = j\Delta t$ with $\Delta t = 1$ month, and $j = 1, 2..., N$ for $N(=240)$ months starting in 1/1988 and ending in 12/2007. A typical $S_{\alpha,g}(t_j)$ is shown in Fig. 4.2.

The logarithmic growth rate $r_{\alpha,g}(t_j)$ is defined by the following:

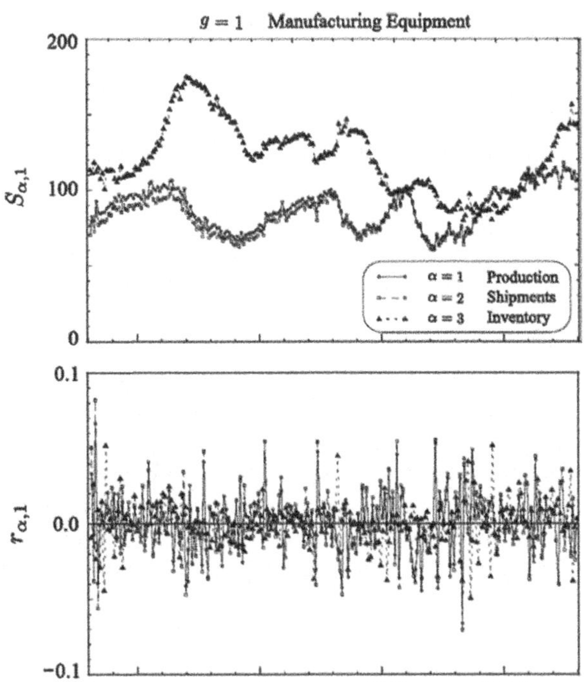

Fig. 4.2 Sample of the IIP data for $g = 1$ (Manufacturing Equipment). *Note* The upper panel shows the indices $S_{\alpha,1}(t_j)$, while the lower panel shows the corresponding growth rates $r_{\alpha,1}(t_j)$ defined by Eq. (4.1). As explained in the legend, $\alpha = 1$ (production) is shown by open circles connected by solid lines, $\alpha = 2$ (shipment) is shown by open squares connected by dashed lines, and $\alpha = 3$ (inventory) is shown by rectangles connected by dotted lines. The horizontal axis shows the months beginning January 1988 through December 2007. *Source* Iyetomi et al. (2011)

$$r_{\alpha,g}(t_j) \equiv \log_{10}\frac{S_{\alpha,g}(t_{j+1})}{S_{\alpha,g}(t_j)} \tag{4.1}$$

where j runs from 1 to $N - 1 = 239$ ($: = N'$). The time series $r_{\alpha,1}(t_j)$ is plotted in the bottom panel of Fig. 4.2.

The normalized growth rate $w_{a,g}$ is then defined by

$$w_{\alpha,g}(t_j) \equiv \frac{r_{\alpha,g}(t_j) - \langle r_{\alpha,g}\rangle_t}{\sigma_{\alpha,g}} \tag{4.2}$$

where $\langle\rangle_t$ denotes the average over time $t_1;\ldots;t_{N'}$ and $\sigma_{\alpha,g}$ is the standard deviation of $r_{a,g}$ over time. Thus, the set $\{w_{\alpha,g}(t_1), w_{\alpha,g}(t_2), \ldots, w_{\alpha,g}(t_{N'})\}$ has an average equal to zero, and the standard deviation is equal to one. The normalization is done because the normalized data are to be used in the subsequent analysis of random matrix theory. We also note that the information revealed by correlation rather than covariance is sufficient for our study of lead–lag relationships among production, shipments and inventories. In the next section, we will examine the cyclical behavior of the normalized growth rate (4.2).

There are two basic problems in the study of business cycles. The first is to identify the periodicity. Second, we must identify common factors causing sectoral comovements. Some economists are skeptical of the stable periodicity of business cycles (Diebold & Rudebusch, 1990). It is certainly true that when we examine the sample path property of a stochastic process, we face difficulties in finding or even clearly defining the periodicity. However, this does not necessarily mean that we cannot usefully explore the periodicity of business cycles. We resort to the standard spectral analysis (see Granger (1966), for example). The result is as follows.

We observe significant power at low frequencies, $T = 60$ months (5 years) and $T = 40$ months (3 years and 6 months). In fact, we can show that the $T = 40$ and $T = 60$ cycles are significant and robust (Iyetomi et al., 2011). The peaks and troughs of Japan's business cycles are officially determined by the Cabinet Office (prior to 2000, the Economic Planning Agency). In the postwar period (October 1951–January 2002), the Japanese economy experienced 12 cycles. The average length of 12 cycles is 50.3 months, with a maximum of 83 months and a minimum of 31 months. The periodicity we have identified, $T = 40$ and $T = 60$ months, is, therefore, quite reasonable.

The dominant factors identified by the random matrix theory

Next, we must identify the significant common factors underlying sectoral comovements. By the common factors, we mean a small number of factors that explain the correlation of the original data as much as possible. We resort to the principal component analysis.

First, we obtain the correlation matrix C as follows:

$$C_{\alpha,g;\beta,l} := \langle w_{\alpha,g} w_{\beta,l} \rangle_t, \tag{4.3}$$

whose diagonal elements are 1 by definition of the normalized growth rate $w_{\alpha,g}$. Because α and β run from 1 to 3 and g runs from 1 to 21, the matrix C has M x M (M = 63) components.

To uncover the unobservable common factors, we examine the eigenvalues $\lambda^{(n)}$ and the corresponding eigenvectors $\boldsymbol{V}^{(n)}$ of matrix C;

$$\boldsymbol{C}\boldsymbol{V}^{(n)} = \lambda^{(n)}\boldsymbol{V}^{(n)}. \tag{4.4}$$

The eigenvalues $\lambda^{(n)}$ satisfy the trace constraint;

$$\sum_{n=1}^{M} \lambda^{(n)} = M. \tag{4.5}$$

The correlation matrix C is written in terms of the eigenvalues and eigenvectors as follows:

$$\boldsymbol{C} = \sum_{n=1}^{M} \lambda^{(n)} \boldsymbol{V}^{(n)} \boldsymbol{V}^{(n)T}. \tag{4.6}$$

The distribution of the eigenvalues $\lambda^{(n)}$ is plotted in Fig. 4.3. We see that there are a few isolated large eigenvalues and that the rest are concentrated in the smaller region. The large eigenvalues and associated eigenvectors correspond to the dominant factors that explain comovements of the fluctuations of different goods.

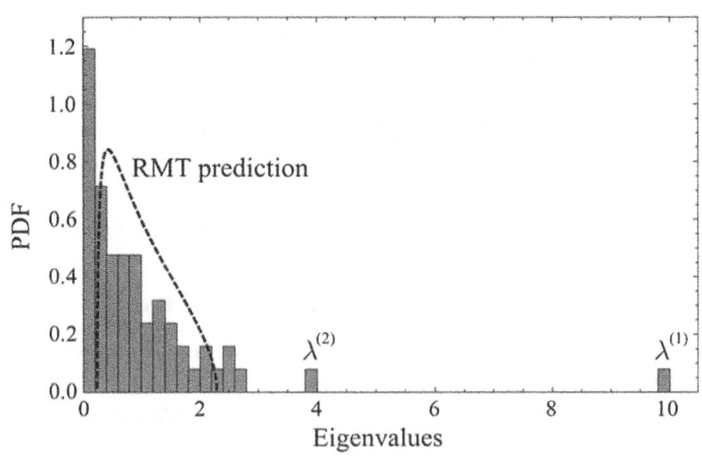

Fig. 4.3 Probability density function (PDF) of the eigenvalues (bars) and random matrix theory (RMT) prediction *q(k)* (dashed curve). *Source* Iyetomi et al. (2011)

The eigenvector corresponding to the largest eigenvalue (which we call "the first eigenvector" hereafter) is the first principal component (PC) that accounts for as much of the correlation as possible in the M-dimensional space. The eigenvector for the second largest eigenvalue ("the second eigenvector") is the second principal component that accounts for as much of the correlation as possible in the subspace orthogonal to the first eigenvector, and so on.

In principal component analysis (PCA), it remains an unresolved problem how to choose significant factors. To solve this problem, many procedures and guidelines have been proposed (Jackson, 1991; Jolliffe, 2002; Peres-Neto et al., 2005). For example, the Kaiser–Guttman rule (Guttman, 1954; Kaiser, 1960), which is very popular in practice, suggests that only factors having eigenvalues greater than one are to be retained. The scree graph named by Cattell (1966) is also widely used. The scree graph is a kind of rank size plot of the eigenvalues of a correlation matrix. In this graph, the abscissa is the rank, and the ordinate is the eigenvalue. The typical shape of the scree graph is a downward-sloping curve, and the slope changes from the "steep" section to the "shallow" section. Accordingly, we can obtain PCs by looking for the point that separates the "steep" and the "shallow". Although scree graphs are used mainly in PCA, Cattell proposes the same method for factor analysis.

The problem of testing the significance of a few dominant factors is also a challenge to exploratory factor analysis (FA) or the so-called approximate dynamic factor model, such as Stock and Watson (1998, 2002). Determining the number of unknown dominant factors has been, in fact, studied in the rapidly growing literature. Bai and Ng (2002), for example, propose criteria for the selection of dominant factors in large dimensions of N and T. They basically use the information criteria in a framework of model selection. Their analyses are, however, done on the asymptotic assumption of large dimensional panels.

How to determine the number of significant components remains an open issue in PCA and FA. Because the multivariate time series of IIP contains uncorrelated shocks and "measurement" errors, the correlation matrix derived from the data is random in nature. It is, therefore, reasonable to compare the properties of the empirical correlation matrix C to a null hypothesis based on a purely random correlation matrix that one would obtain with panel data of random variables of the same size as the IIP data. The null hypothesis, in which variables are totally uncorrelated, enables us to filter C for significant components. If the random data were infinite, their correlation matrix would become the unit matrix. However, we must stress that the random correlation matrix for the null hypothesis is not trivial because the length of the available data is finite.

The *random matrix theory* (RMT) arguably provides us with the best filtering procedure of C currently available. Thanks to the RMT, we can use a wide variety of analytic formulas with regard to random matrices, including the probability density function (PDF) for the eigenvalues of the random correlation matrix. The RMT has made great success in the physical sciences and has quickly found applications to various disciplines. We note that the RMT has been applied to finance for a decade (see the textbook by Bouchaud and Potters (2000) and references therein) and macroeconomics very recently (Ormerod, 2008). An interested reader is referred to Iyetomi (2011).

Using RMT, we conclude that the two largest eigenvalues

$$\lambda^{(1)} \approx 9.95, \lambda^{(2)} \approx 3.82,$$

are significantly larger than the upper bound λ_+, hence reflecting the nonrandom nature of the data.

We have demonstrated that the two largest eigenvalues are statistically meaningful. In Fig. 4.4, the components of the eigenvectors $V^{(1,2)}$ are shown for production, shipments, and inventory by type of goods. Taking advantage of this information revealed by disaggregated data, we can identify two eigenvectors as follows.

We observe that the first eigenvector $V^{(1)}$ captures the comovements of production and shipments. Thus, the first dominant factor corresponding to the largest eigenvalue is best interpreted as "aggregate demand" because, by definition, it moves both shipments and production in all sectors. This is Keynes's principle of effective demand. The textbook multiplier analysis depends, of course, on the assumption that changes in shipments induce almost equal changes in production.

It is noteworthy that in 15 out of 21 cases, responses of inventory stock are negative, suggesting depletion of inventory stock in response to the first dominant factor $V^{(1)}$, which we interpret as the aggregate demand. When a firm's production is subject to increasing marginal cost and/or production takes time, there is a buffer-stock motive for inventory. The buffer stock motive produces a negative correlation between inventory investment and sales when demand exogenously changes (Hornstein, 1998).

The second dominant factor, $V^{(2)}$, on the other hand, captures the comovements of production and inventory. Its effect on shipments is relatively small. Metzler's (1941) classical work demonstrates that production is made not only for the purpose of meeting the current sales or shipments but also for maintaining the firm's desired level of inventory stocks. The firm's desired level of inventory stocks depends on anticipations of future demand or sales. That is, inventory investment becomes, in part, "speculative". In fact, based on a robust stylized fact that the variance of production is greater than that of shipments, Blinder and Maccini (1991) and others have demonstrated that the role of inventory stock is not confined to production smoothing. Rather, they show that inventories apparently destabilize output at the macro level. Whether originating from the Metzlerian inventory accelerator (the idea that desired inventory stocks rise with sales) or from the (S, s) inventory strategy at the micro level, inventory and production tend to move together. Our second dominant factor, $V^{(2)}$, appears to capture this channel. For convenience, we call it "inventory adjustments".

Whether direct or indirect, $V^{(1)}$ and $V^{(2)}$ are basically demand factors. It is difficult to interpret $V^{(1)}$ and $V^{(2)}$ as "productivity shocks" for the following reason. Changes in total factor productivity (TFP) emphasized by real business cycle theory

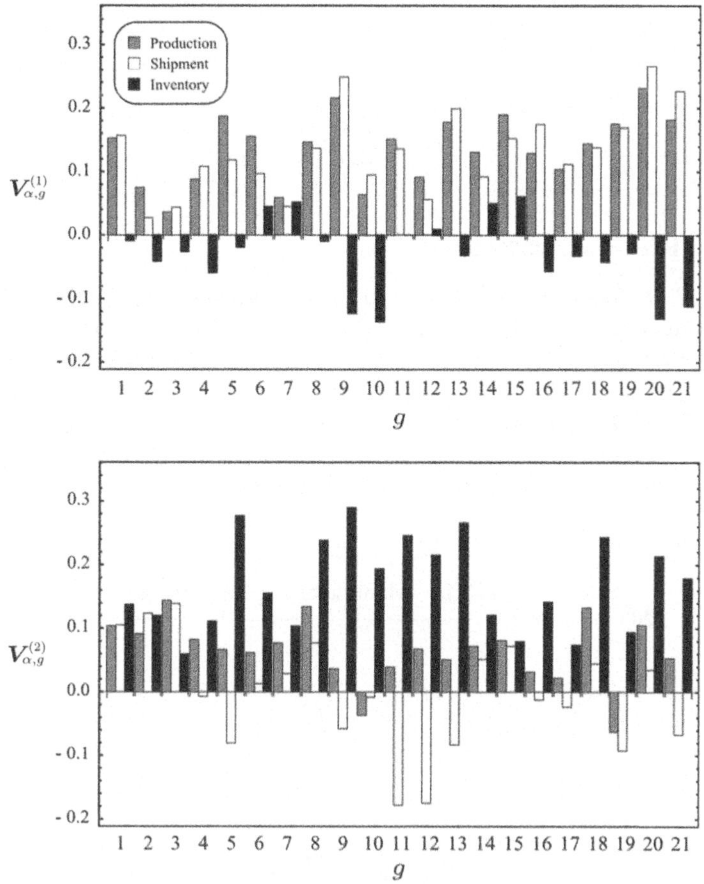

Fig. 4.4 Components of the two dominant eigenvectors, $V^{(1,2)}$. *Note* The horizontal axis shows the 21 goods ($g = 1, 2..., 21$). The gray bars are the production ($a = 1$), the white bars are the shipments ($a = 2$), and the black bars are the inventory ($a = 3$). *Source* Iyetomi et al. (2011)

such as Kydland and Prescott (1982) bring changes in marginal cost in production. To minimize costs, the firm increases (decreases) production and accumulates (reduces) inventories during times when marginal cost is low (high). Thus, not only final demand (consumption/investment) usually analyzed by RBC but also inventory investment moves together with production when productivity changes (Hornstein, 1998). In other words, productivity shocks, in particular temporary shocks, must move all the variables, namely, production, shipments, and inventory, in tandem. Figure 4.4 clearly shows that $V^{(1)}$ captures the comovements of production and shipments, while $V^{(2)}$ captures the comovements of production and inventory, leaving the third variable relatively unaffected in both cases. The two dominant factors, therefore, do not represent productivity shocks.

We note that we can decompose the normalized growth rate $w_{\alpha,g}(t_j)$ on the basis of the eigenvectors $V^{(n)}$;

$$w_{\alpha,g}(t_j) = \sum_{n=1}^{M} a_n(t_j) V_{\alpha,g}^{(n)}. \tag{4.7}$$

The Great Recession, 2008–2009

The 2008–2009 financial crisis triggered by the bursting of housing bubbles in the US immediately spread all over the world and caused great damage to the global economy, reminiscent of the Great Depression. Based on the analyses above, we update the IIP data to cover the period from January 2008 ($j = 241$) through June 2009 ($j = 258$) and perform an out-of-sample test.

The extended data shown in Fig. 4.5 demonstrate a dramatic drop in production in Japan. Panel (a) of Fig. 4.6 shows the volatility $P(t)$ of the IIP defined by

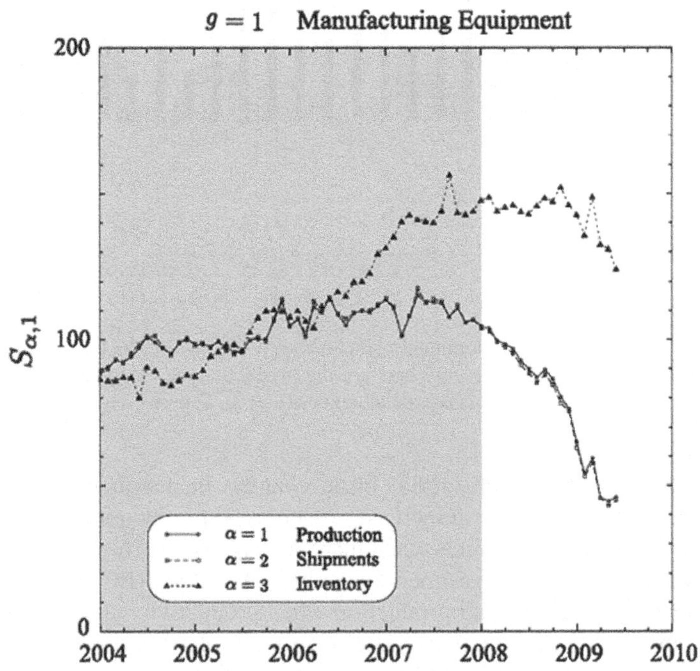

Fig. 4.5 The 2008–2009 recession. *Note* The IIP data for $g = 1$ (Manufacturing Equipment) are extended to encompass the 2008–2009 economic crisis. The gray shaded zone depicts the in-sample period used in the analyses production and shipments almost overlap with each other and are hard to distinguish. *Source* Iyetomi et al. (2011)

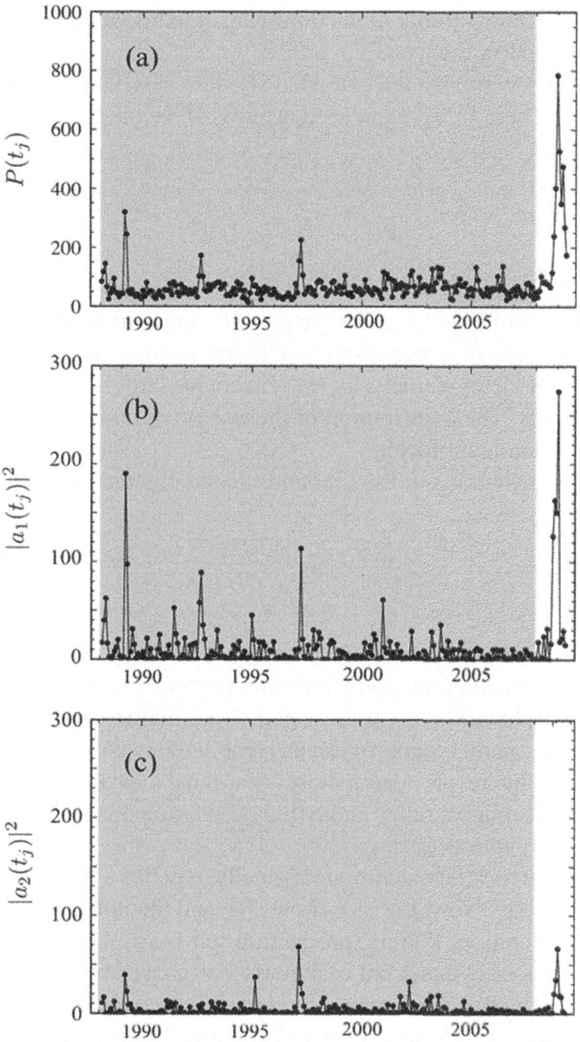

Fig. 4.6 Decomposition of volatility of the extended IIP data. *Note* (a) Total volatility. (b) The partial contribution of the first eigenmode to the total volatility. (c) The partial contribution of the second eigenmode. The gray shaded zone depicts the in-sample period used for determining the dominant modes. *Source* Iyetomi et al. (2011)

$$P(t) := \sum_{\alpha=1}^{3} \sum_{g=1}^{21} \left| w_{\alpha,g}(t) \right|^2. \tag{4.8}$$

We observe the exceptionally large amplification of $P(t)$ during the 2008–2009 economic crisis. This situation, therefore, provides us with a perfect opportunity for an out-of-sample test for validity of the dominant eigenmodes that we have identified using the random matrix theory.

To this end, we decompose the extended IIP data based on the exact same eigenvectors as in (4.7). Then, $P(t)$ is expressed in terms of the mode amplitude coefficients $a_n(t)$ as follows:

$$P(t) = \sum_{n=1}^{63} |a_n(t)|^2. \tag{4.9}$$

Panels (b) and (c) of Fig. 4.6 show the partial contributions of the largest and the second largest modes to $P(t)$, respectively. Plainly, an unprecedented rise of volatility of the first dominant factor, namely, $|a_1(t)|^2$, accounts largely for a rise of $P(t)$ during the 2008–2009 crisis. The contribution of the second dominant factor is not as great as that of the first dominant factor.

The relative contributions of the first and second eigenmodes defined by

$$\pi_n(t) := \frac{|a_n(t)|^2}{P(t)}. \tag{4.10}$$

are shown in Fig. 4.7. The total sum of $\pi_n(t)$ with respect to n is normalized to unity. The first dominant eigenmode accounts for almost half of the total power of fluctuations during the 2008–2009 recession. Note that the relative contribution of the first eigenmode, namely, approximately one half, is comparable to that for the in-sample period. This result demonstrates that our factor model holds up over the 2008–2009 crisis although a fall of industrial production during the period, as shown in Fig. 4.5, was truly unprecedented.

We have seen that the first eigenmode basically explains a fall of industrial production during 2008–2009. Now, Fig. 4.8 shows IIP and the index of exports during the period under investigation. Plainly, production fell in response to a sudden decline in exports. We emphasize that a fall of exports was caused by a sudden shrinkage of world trade due to the financial crisis accompanied by the global recession and that it was basically exogenous to the Japanese economy. It was certainly independent of technological shocks. By analyzing the components of eigenvectors (Fig. 4.4), we identified the first dominant factor as the aggregate demand. To the extent that we identify a sudden fall of exports, an exogenous negative aggregate demand shock, as the basic cause of the 2008–2009 recession, the dominance of the first eigenmode provides us with another justification for our interpretation of it as the aggregate demand factor. This confirms once again the validity of the Old Keynesian View (Tobin, 1993).

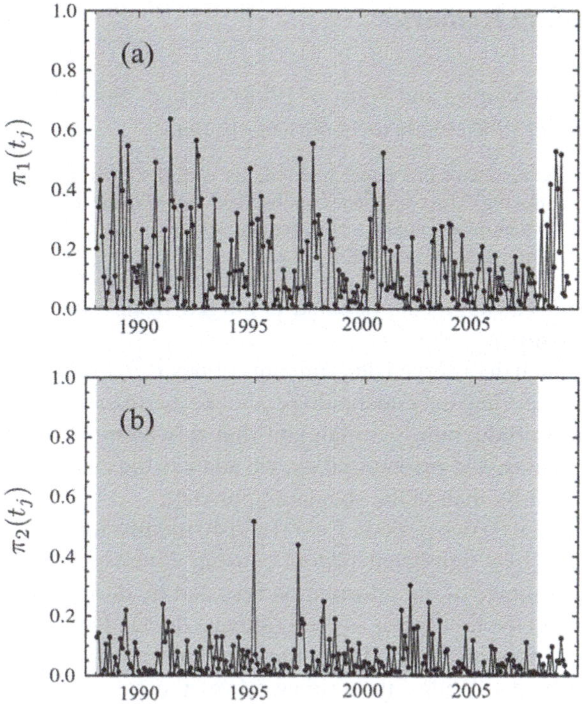

Fig. 4.7 Relative contributions of (a) the largest mode and (b) the second largest mode. *Note* Panels (a) and (b) correspond to panels (b) and (c) in Fig. 4.6, respectively. *Source* Iyetomi et al. (2011)

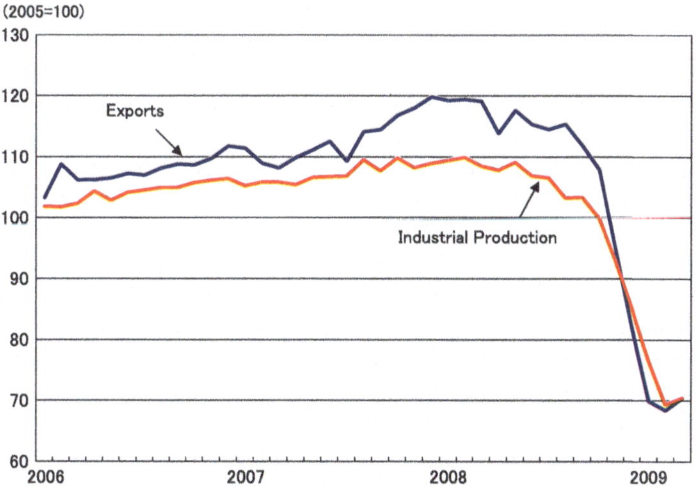

Fig. 4.8 Indices of exports and industrial production, normalized to 100 for 2005. *Source* Cabinet Office, Ministry of Economy, Trade and Industry

4.3 Concluding Remarks

In his comments on Shapiro and Watson (1988) entitled "Sources of Business Cycle Fluctuations," Hall (1988) made the following remark:

> All told, I find the results of this paper harmonious with the emerging middle ground of macroeconomic thinking. There are major unobserved determinants of output and employment operating at business-cycle and lower frequencies. Some are technological, some are financial, some are monetary (Hall, 1988, 151).

In his remark, *real* aggregate demand is purged from a list of determinants of output and employment.

To determine such unobserved determinants of business cycles, the standard identifying restriction is that only technology shocks permanently affect output. As discussed in the Introduction, we maintain that this assumption is wrong. Using production, shipments, and inventory data, we analyze the causes of business cycles more straightforwardly than in the standard approach.

We have identified (i) two periods $T = 60$ and 40 months for the Japanese business cycles and (ii) the two significant factors causing comovements of IIP across 21 goods. Taking advantage of the information revealed by disaggregated data, we can reasonably interpret the first factor as the aggregate demand and the second factor as intentional inventory adjustments. It is difficult to interpret the two dominant factors as productivity shocks. Within the framework of the standard RBC, productivity shocks are expected to affect production, shipments, and inventory together. The two dominant factors we have identified do not fulfill this condition (Fig. 4.4).

We have also found that for the part of shipments and production explained by two factors, shipments lead production by a few months. This lead–lag relationship between shipments and production caused by inventory adjustment is not clearly seen in the original data. However, inventory adjustments that generate this lead–lag relationship between shipments and production are an important part of "business cycles" defined by two dominant factors.

Furthermore, out-of-sample analysis of the 2008–2009 recession demonstrates that the first dominant factor largely explains this major episode. We know that it was basically caused by a sudden fall of exports beginning the fourth quarter of the year 2008 (Fig. 4.8). Therefore, to the extent that exports can be reasonably regarded as exogenous real aggregate demand shocks to the Japanese economy, we can identify the first dominant factor as the real aggregate demand. All these findings suggest that the major cause of business cycles is real demand shocks accompanied by significant inventory adjustments.

It is important to recognize that this conclusion of ours does not necessarily mean that technology shocks are unimportant. In contrast, it is very plausible that a major driving force of fixed investment is technology shocks. Furthermore, one might argue that technology shocks are also a driving force of consumption by way of introduction of new products (see Aoki and Yoshikawa model in Chap. 5). The point is that technology shocks affect the macroeconomy by way of changes in aggregate demand, notably investment, *in the short run*. For the purpose of studying business cycles, it is wrong to consider technology shocks as *direct* shifts of the production function.

The conclusion we drew from our analyses of the Japanese industrial production data is clear. Studying whether the same results hold up for other economies is obviously an important research agenda. We believe that the methods we employed in the present paper can be usefully applied to the study of business cycles in other economies.

Finally, we have identified 40–60 months "cycles" of aggregate demand. Keynes made the following remarks:

> If we examine the details of any actual instance of the trade cycle, we shall find that it is highly complex and that every element in our analysis will be required for its complete explanation. In particular we shall find that fluctuations in the propensity to consume, in the state of liquidity-preference, and in the marginal efficiency of capital have all played a part. But I suggest that the essential character of the trade cycle and, especially, the regularity of time-sequence and of duration which justifies us in calling it a *cycle*, is mainly due to the way in which the marginal efficiency of capital fluctuates. The trade cycle is best regarded, I think, as being occasioned by a cyclical change in the marginal efficiency of capital, though complicated and often aggravated by associated changes in the other significant short-period variables of the economic system (Keynes, 1936, 313).

The mechanism of aggregate fluctuations with the regularity of time-sequence and of duration, which justifies us in calling them "cycle", was once fully analyzed; the multiplier–accelerator models of Samuelson (1939), Hicks (1950) and Goodwin (1951) provide us with a useful framework for analyzing such "cycles" of aggregate demand.

Chapter 5
Demand Saturation and Economic Growth —Where Keynes and Schumpeter Meet

How does the economy grow? Economic growth, together with the business cycle, is the central theme in macroeconomics.

In 2017, the per capita GDP of Luxembourg, the world richest country, was a hundred thousand U.S. dollars. Ranked 8th, U.S. per capita income was sixty thousand dollars, 19th Germany forty-five thousand dollars, following France and the UK, and 25th Japan's income was thirty-eight thousand dollars. On the other end, ranked 191th in the IMF statistics, the per capita income of South Sudan was only 228 dollars. Countries in central Africa are poor, with average incomes of 300–400 dollars. Ranked 180th, the average income of Afghanistan was 600 dollars, one hundredth of the U.S. income. The fundamental reason for such differences of incomes on a scale of a hundred to one is different growth rates.

China's GDP is currently 12 trillion dollars next only to the U.S. whose GDP is 19 trillion dollars. Within ten years, however, China is expected to catch up with the U.S. and become the world's largest economy. High economic growth since the 1978 Reform has made it possible.

The concept of economic growth, however, has not always existed in human history. It was born in the 18th century when modern capitalism took off. Arguably, economics in its present form is also a byproduct of this new age. Adam Smith's *An Inquiry into the Nature and Causes of the Wealth of Nations* published in 1776 is nothing but a kind of growth theory.

In this chapter, we first review the modern theory of economic growth "old" and "new". We pay special attention to Arthur Lewis's (1915–1991) theory of development and explain that it well accounts for the growth experiences of the Japanese economy. Then, we argue that the most important factor restraining the growth of the advanced economy is the *saturation of demand*. It is one of the few iron laws in economics. Conversely, the engine for growth is innovations generating new major products that command high growth of demand. We present a model of economic growth based on demand saturation and product innovations. Finally, we briefly consider the relationship between population and economic growth.

H. Yoshikawa, *Reconstruction of Macroeconomics: Methods of Statistical Physics, and Keynes' Principle of Effective Demand*, Advances in Japanese Business and Economics 3, https://doi.org/10.1007/978-981-19-5264-7_5

5.1 Growth Theory

Keynes's General Theory takes stock of capital and labor supply given. Investment plays the central role only as a part of aggregate demand. However, net investment plainly increases the stock of capital on the supply side. Its role is, therefore, dual; investment changes not only the level of aggregate demand but also the level of capital stock in the near future. Accumulated capital stock must be productively used, but capacity utilization depends on aggregate demand, which in turn depends crucially on investment. Is the relation between demand and supply harmonious in the process of economic growth? This is the problem analyzed by Roy Harrod (1939). He showed that the harmonious growth path that he called "the warranted rate of growth" is s/v. Here, s is the saving rate, and v is the capital coefficient, namely, the stock of capital K required for producing a unit of output Y, or K/Y. By a slightly different route, Domer (1946) also reached the same conclusion.

Harrod also showed that the warranted growth path is similar to the "knife edge", meaning that it is unstable. That is, if the actual growth rate is not exactly equal to s/v, it further diverges from rather than returning to s/v. Harrod succeeded Keynes's vision that the *laissez faire* economy is intrinsically unstable.

In contrast to Harrod, Solow (1956) demonstrated that the balanced growth path s/v is stable using the neoclassical production function $Y = (K, L)$. In the neoclassical production function, the substitution of two factors of production, capital K and labor L, is allowed, making it possible for the same amount of output Y to be produced with different combinations of capital and labor. Harrod's result of instability comes from the rigid assumption that s and v are both given constants. In Solow's neoclassical growth model, v is variable as a result of the substitution of K and L.

Solow evidently had in mind decades long or even a century long process of economic growth. Then, instability seemed odd. On Harrod/Domer models, Solow made the following remark in his Nobel Lecture in 1987:

> Nevertheless, it seemed to me that the story told by these models felt wrong. An expedition from Mars arriving on Earth having read this literature would have expected to find only the wreckage of a capitalism that had shaken itself to pieces long ago. Economic history was indeed a record of fluctuations as well as growth, but most business cycles seemed to be self-limiting. Sustained, though disturbed, growth was not a rarity (Solow, 2000, x).

Harrod and Solow analyzed growth in different time spans. For example, Harrod considers a firm's expectations of demand and capacity utilization in his analysis. Firms' expectations of demand make sense only in the relatively short period, i.e., two to three years. On the other hand, factor substitution in Solow's model plays an important role in longer periods, perhaps longer than five years. The neoclassical growth model is meant to describe the process of economic growth over a very long period, say several decades. Solow (2005) actually dwells on the issue and concludes that "the natural habitat of growth-theoretic explanations is time spans of 25 to perhaps 40 or 50 years." As such, neoclassical growth theory is an all-time winner in the sense that it provides a useful accounting framework for the analysis of economic growth.

Soon after, based on his own model, Solow (1957) analyzed the growth of the U.S. economy empirically and demonstrated that more than 90% of growth resulted from "technical progress" (later called total factor productivity = TFP) rather than the accumulation of capital. To be concrete, he estimated the following production function:

$$Y = AF(K, L) \tag{5.1}$$

Here, A is the index of "technology" or TFP. Unfortunately, we do not have data on A. From "growth accounting" to this day, A is estimated as a "residual", namely, a part of Y that cannot be explained by K and L.

We may note that there is actually a discrepancy between theory and standard practice in empirical works. Technical progress A, which is outside the production function as in Eq. (5.1), is *Hicks neutral*. However, Uzawa (1961) demonstrates that nice properties of neoclassical growth theory, including the existence and stability of steady growth paths, hold true only in the case of *Harrod's neutral* or labor-augmenting technical progress, namely, $Y = F(K, AL)$, not in the case of Hicks' neutral technical progress. Harrod neutral and Hicks neutral technical progress happily coincide only if the production function is of the Cobb–Douglas type. More generally, Sato and Beckman (1968) clarify the relation between the factor augmenting technical progress and the elasticity of substitution. Sato and Morita (2009) convincingly show that the factor augmenting (biased) technical progress is more likely than TFP.

In any case, A is a black box. Then, it may be natural that economists seek to "endogenize" technology or A. The attempts bore the huge literature of "endogenous growth theory"; Romer (1986), Lucas (1988), Grossman and Helpman (1991), Aghion and Howitt (1992). In Solow's "old" model, the growth rate ultimately depends on the exogenously given growth of the population and technology. In the "new" endogenous growth models, the growth is sustained thanks to endogenously generated technical progress.

These models are all "microfounded" resting on representative agent assumptions. The representative Ramsey consumer maximizes the discounted sum of utility, while the representative firm maximizes profits being engaged not only in ordinary production of goods/services but also in research and development (R&D). The criticism of the representative agent approach made in Chap. 1 certainly applies to these models.

Beyond that, the endogenous growth models all rest on the crucial assumption that the key parameter—call it α—is exactly one. If α is less than one, the return on R&D is diminishing, and the growth rate asymptotically approaches zero unless there is no exogenously growing factor such as population. On the other hand, if α is larger than one, the growth rate explodes; namely, it becomes infinite in finite time. Thus, positive growth is sustained only if α is exactly equal to one. The assumption lacks robustness—of measure zero on parameter space—and is untenable (Solow, 1994, 2000).

There is no doubt that technical progress, or more broadly, innovation, plays a decisive role in economic growth. An important part of innovations is done by profit-motivated private firms. However, the government and many nonprofit organizations such as universities, also play an important role (Postan, 1971). It is useless to incorporate R&D into growth models based on representative agents.

5.2 Lewis Model

Arthur Lewis's (1954) theory is usually taken as a model of development that applies to developing countries. True, the title of his original paper is "Economic Development with Unlimited Supplies of Labour." In advanced economies, labor supply is certainly not unlimited. However, in Lewis' model, labor supply is not physically unlimited. The point is that the economy is dual in that the marginal products of labor differ across two sectors. As such, the concept applies usefully to the analysis of the process of growth of advanced economies. The difference between "development" and "growth" is of little significance. Kaldor (1966), for example, in his inaugural lecture at Cambridge entitled "Causes of the Slow Rate of Economic Growth of the United Kingdom", compared the share of agriculture for twelve advanced economies and concluded that the high economic growth of Japan and Italy during the 1950s–60s was due to the presence of the agricultural sector in these countries. Kindleberger (1967) also advanced a similar explanation for Europe's postwar growth. Temin (2017) attempts to explain the U.S. inequality in the 2000s by the Lewis model. As we will see in what follows, the model is a very useful framework for analyzing the growth of advanced economies.

The economy consists of two sectors, namely, the "traditional agriculture sector" and the "modern industrial sector". The level of real wages is determined at the "minimum subsistence" level in the traditional sector, where the marginal products of labor are lower than the "minimum subsistence level". The "minimum subsistence level" must mean the socially acceptable level and is not necessarily physiologically determined. The modern industrial sector can employ as many workers as it wants at this given level of real wages. Because the marginal products of labor in the industrial sector are higher than real wages, it can enjoy high profits exceeding the marginal products of capital. Thus, the modern sector takes advantage of excess profits and carries out high investment to accumulate capital.

Responding to the growth of the modern industrial sector, labor flows out of traditional agricultural villages to cities.[1] As labor decreases in the traditional sector, the marginal products there gradually rise, exceed the exogenously given minimum subsistence level, and finally catch up with the marginal products of labor in the

[1] The migration of labor from the traditional agricultural sector to industrial cities in the process of economic growth was long noticed by economists. For example, Cassel (1932) emphasized the positive relation between industrialization and migration based on the Swedish experiences. Myrdal (1933) provides empirical support for Cassel's proposition.

modern sector. Then, the marginal products in the two sectors are equated, and the economy reaches neoclassical equilibrium. It is called “the turning point” in the literature.

Japan is one of the few countries that crossed this turning point in the 20th century. A group of Japanese economists led by Kazushi Ohkawa bore huge literature on century-long growth of the Japanese economy from the late 19th century to the post World War II (WWII) period; see Ohkawa and Rosovsky (1973), Minami (1970) and references cited therein.

The Lewis model is a good model, but there are several serious problems.

(1) In the model, it is assumed that the economy eventually reaches the “turning point” and that thereafter, the economy is in neoclassical equilibrium. In particular, the marginal products of labor are equal in all sectors. However, it is shown in Chap. 3 that the dispersion of labor productivity never ceases even in the fully developed economy. True, historically, in every country, the greatest gap was between the traditional agricultural sector and the modern industrial sector. At the “turning point”, the traditional sector with “excess labor supply” disappears. However, that does not mean that the economy reaches the neoclassical equilibrium where the marginal products are equal in all the sectors/firms. We do not have a unique marginal product of labor but rather a distribution of productivity. The economy is not dual but actually consists of multiple sectors with different levels of productivity. In this sense, the modern economy always carries *under-employment* as well as unemployment. In Chap. 3, we showed that it provides the basis for Keynes's principle of effective demand.
(2) The second problem with the Lewis model is that the growth of the modern industrial sector comes solely from high profits sustained by low real wages. There is no analysis of demand for goods and services produced in the modern sector. Low real wages may not be advantageous to vigorous domestic demand. Here, it is extremely useful to recall the high growth of the Japanese economy during the 1950s and 1960s.

Japan's Postwar High Growth

Underlying Japan's rapid growth was the people's ardent desire to modernize their lifestyles—which meant a consuming drive to “Americanize” their lives and catch up with the United States. Drawing upon this latent demand, firms embark upon technological innovation and the capital investment needed to incorporate the new technologies into their production facilities.

Technological innovation encouraged investment in new plants and equipment, which then enabled firms to achieve an economy of scale, both reducing production costs and improving product quality. This ambitious investment in steel and other upstream industries then made it possible for a wide range of downstream industries to achieve cost reductions and quality enhancements as well. Typical examples were downstream consumer durables such as washing machines and television sets, where costs and hence prices could he slashed sufficiently to put products within reach of the vast latent demand. When prices went down and incomes went up, the result was a wildfire-like diffusion of consumer durables. Because this frenzied rush to buy consumer durables generated additional demand for high-grade metals and plastics, it induced additional innovation and investment in these material industries.

The higher wages that were paid to workers in the more productive urban manufacturing industries drew rural residents, especially young people, to the cities, with the collateral result that there was a sharp rise in the number of people living alone or in nuclear families. The increase in the number of households was undoubtedly one result of Japan's rapid growth, but it was also a driving factor in creating increased demand for goods and services. As these young people left their extended-family households in the countryside and set up new households in the city, they inevitably generated new demand for consumer durables, electricity, and more. The movement from farm to city and the consequent increase in the number of households were the results of economic growth, but at the same time also the causes of high economic growth. Thus, the whole mechanism of high growth was a circle, as shown in Fig. 5.1.

How were wage hikes and expanded capital investment compatible? In the Lewis model, it is low real wages that drive high investment in the modern industrial sector. Yes, low wages were advantageous when Japanese growth was driven by exports primarily of textiles before WWII. However, Japan's postwar growth was fundamentally different from the prewar growth in that it was basically led by domestic demand.

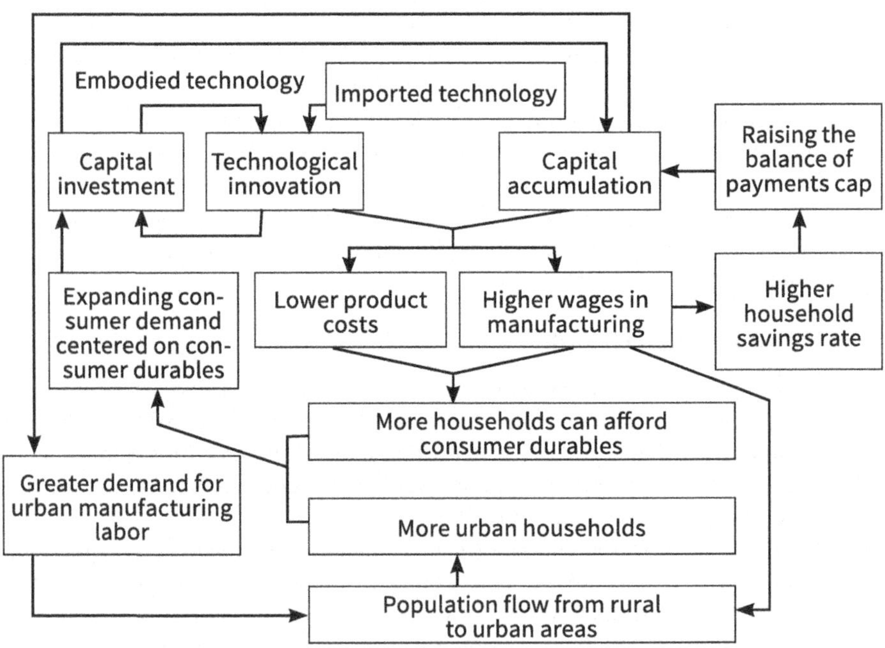

Fig. 5.1 The circulatory mechanism of rapid growth. *Source* Yoshikawa (2021)

Here, high wages were actually advantageous to generating vigorous demand, particularly demand for consumer durables such as washing machines and television sets.

This is not, of course, to say "the higher the better" for wages. However, it is to argue that vigorous capital investment in line with economic growth should result in improved productivity due to newer technologies and economies of scale and that these productivity enhancements should then give firms the leeway to raise wages commensurately. Unlike the prewar Japanese economy, which saw numerous bouts of wage stagnation, the rapid-growth economy saw higher wages as a definite plus factor for sustaining growth led by domestic demand.

At the beginning of the 1970s, the diffusion of then-available consumer durables was largely complete (Fig. 5.2), and new capital investment could not be justified by stronger domestic demand. This brought the rapid growth to an end. As we will shortly see, the saturation of demand is the key factor for economic growth.

5.3 Saturation of Demand

The demand for existing goods and services always becomes saturated. This reality, which should be deservedly called a law, is a fundamental factor that restricts the growth of advanced economies. Initially, demand—and with it, production—shows high growth, but the growth inevitably slows eventually. In extreme cases, beyond suffering a slowdown in demand, some goods and services disappear by what Schumpeter calls "creative destruction". Coal for heating is a typical example.

In this way, the demand for goods and services, and consequently production, initially increases over time. Nevertheless, the growth rate eventually slows to zero, ultimately reaching a ceiling. This means that the growth path is not exponential, as is typically assumed in economic theory, but rather follows an S-shaped *logistic curve*. The growth rate of a logistic curve accelerates at first, but it eventually reaches the inflection point and then declines toward zero, thereby forming an S-shape overall.

Regarding demand saturation, a great deal of empirical research has been done concerning the "life cycle" of growth. For example, engineers J.C. Fisher and R.H. Pry (1971) proposed a model in which the substitution of new goods for old goods plays a key role. In this model, the growth of newly released goods follows a logistic curve. Fisher and Pry applied this model to natural rubber, synthetic rubber, and a variety of manufactured products and demonstrated that, in fact, the growth of many goods follows a logistic curve. Figure 5.3 applies a standardized logistic curve with a ceiling to various goods that appeared in the U.S. over a period of almost 100 years. At a glance, one can see that the growth of many goods follows this pattern.

For consumer durables such as refrigerators, TVs, air conditioners and automatic coffee makers, Bass (1969) demonstrates that sales of many consumer durables show exponential growth in the initial phase, but the growth eventually decelerates. He explains this empirical fact, the S-shaped curve, by a behavioral model in which there are different types of consumers. "Innovators" decide to purchase a new

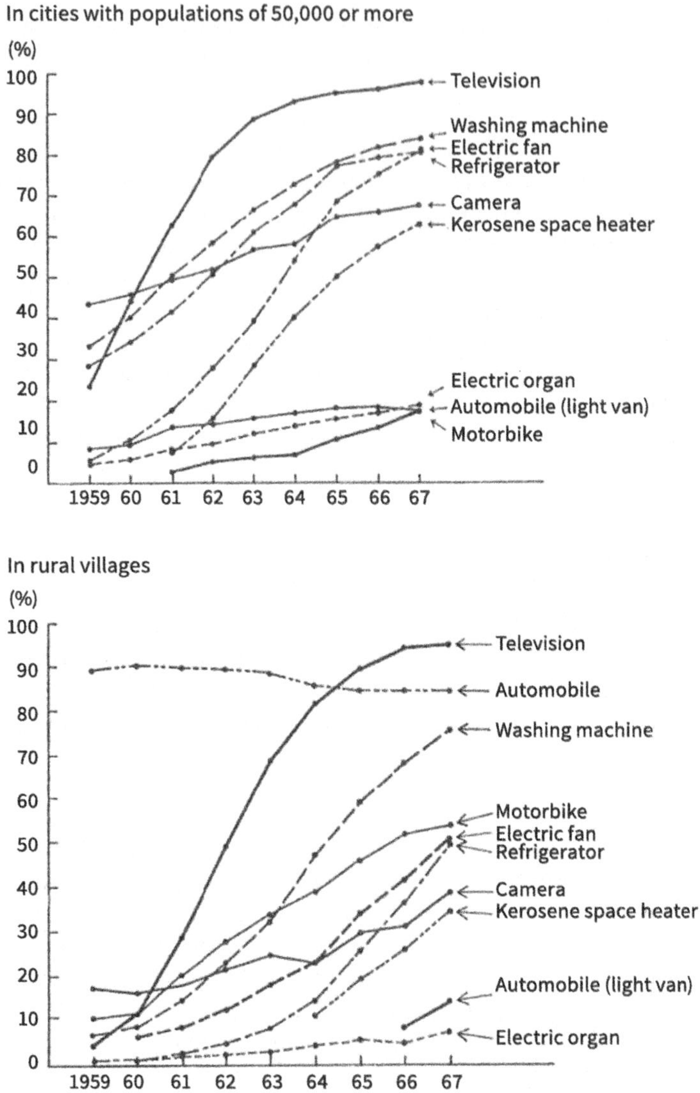

Fig. 5.2 Diffusion of consumer durables, Urban versus Rural, 1959–67. *Source* Economic planning agency, trends of consumption and savings, 1966

product independently of the decisions of other individuals, while "imitators'" decisions depend on the number of previous buyers. Bass analyses the U.S. market. The Japanese experience, as shown in Fig. 5.2, is in good accordance with his analysis.

What Fisher/Pry and Bass investigated is primarily saturation of demand in the manufacturing industry, but economists have been actually noting the saturation of demand for a long time. The most famous of these observations is *Engel's Law*,

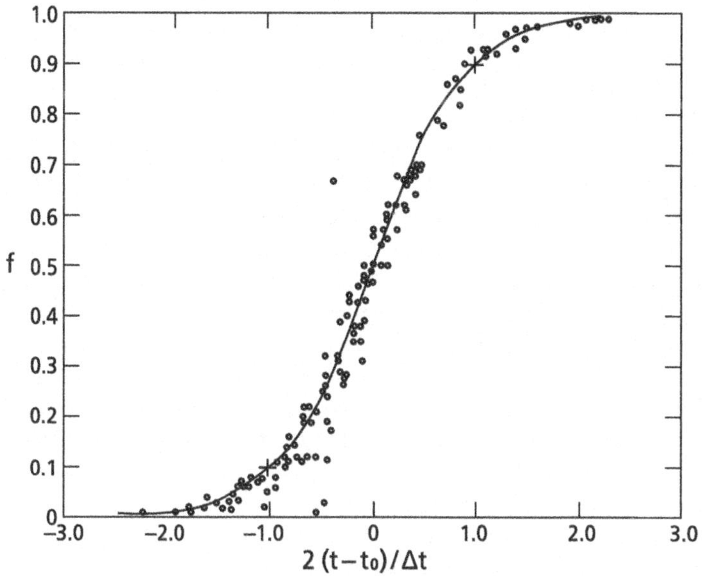

Fig. 5.3 Logistic curve. *Source* Fisher and Pry (1971, 87), Fig. 9b

the observation made by the German statistician Ernst Engel (1821–96) using a household income survey in Belgium in 1895.

This well-known maxim states that the wealthier the household is, the smaller the percentage of consumption expenditures spent on food (the Engel coefficient). This holds true all times and in all countries. When the consumption behavior of many households in a given society is investigated for a given year—which is what Engel did—Engel's Law holds true. Furthermore, on a macroeconomic level, when two countries with different levels of per capita income (such as Japan and China) are compared, we can see the relationship discovered by Engel: Japan, which has a higher income level than China, has a lower Engel coefficient than China does. Similarly, when conditions within Japan in the nineteenth century are compared with the modern day, Engel's Law also holds true. In this way, Engel's Law is one of the few "laws" of economics that is truly deserving of that title.

What this law means is that the demand for food becomes saturated. This is because of the simple fact that there are physiological limits to the food demand of one human being. Regardless of how expensive meals—even fancy French cuisine—may be, the demand is ultimately constrained by the physical limits of the stomach!

Because the demand for food and thus for agriculture—the industry that produces food—becomes saturated, economic growth beyond that demand is led by industries other than agriculture, that is, by manufacturing and services. As a result, if per capita income increases, the market share of agriculture inevitably declines. In other words, Engel's coefficient declines. As mentioned above, when the demand for food becomes saturated, Engel's Law is upheld. The law of demand saturation that Engel discovered concerning food clearly applies to goods other than food. As Fisher/Pry, Bass and others discovered, demand for any good will inevitably become saturated.

This most important fact for economics has been noted by great economists in the past. In April 1930, David Robertson, a friend of Keynes who later became the professor of economics at Cambridge after Pigou, expressed his own thoughts about the ongoing Great Depression at the MacMillan Committee established by the British government. Robertson cited "the gluttability of wants" (i.e., the saturation of demand) as the main cause of the Great Depression. He then said that this problem is "at once the most fundamental, the most difficult to analyze, and the most difficult to counteract." When the demand for large numbers of goods and services is saturated, the economy falls into a major recession. To overcome this, he said, there is no choice but "the perpetual stimulation of new wants." Robertson, who had a touch for irony, continued in his somewhat cynical style as follows. "Certainly the nation that has practiced this immoral art with most assiduity is that which has longest postponed the disease of industrial depression."

Robertson uses the verb "postpone" rather than "solve" in the belief that even if the stimulation of wants continues, in the end, the economy cannot overcome the saturation of demand and thereby avoid the eventual onset of economic depression. Several years after Robertson's testimony, Keynes, in his *General Theory,* also wrote about the saturation of demand:

> Ancient Egypt was doubly fortunate, and doubtless owed to this its fabled wealth, in that it possessed two activities, namely, pyramid-building as well as the search for the precious metals, the fruits of which, since they could not serve the needs of man by being consumed, did not stale with abundance. The Middle Ages built cathedrals and sang dirges. Two pyramids, two masses for the dead, are twice as good as one; but not so two railways from London to York. Thus, we are so sensible, have schooled ourselves to so close a semblance of prudent financiers, taking careful thought before we add to the "financial" burdens of posterity by building them houses to live in, that we have no such easy escape from the sufferings of unemployment (Keynes, 1936, 131).

The marginal benefits generated by the pyramids of ancient Egypt and the cathedrals of the Middle Ages do not decrease, regardless of how many are built. Consequently, the demand is never saturated. However, in the economies of contemporary industrialized nations, the demand for existing goods and services always becomes saturated. In this way, the economy is chronically troubled by demand deficiency. Even if such goods and services are produced, they will not sell (i.e., there is no demand), so firms do not produce. This leads to unemployment. What lies behind Keynesian economics or the principle of effective demand is the fact that the demand for existing goods and services inevitably becomes saturated.

If the demand for existing goods and services becomes saturated, then as long as the list of goods and services does not change, the growth of the economy overall will converge toward zero. In a mature economy, where many goods and services have become widespread, there is always downward pressure on economic growth. While this is only natural, the source of growth in the economies of such industrialized economies is the birth of new goods and services that enjoy high demand growth; in other words, "product innovation."

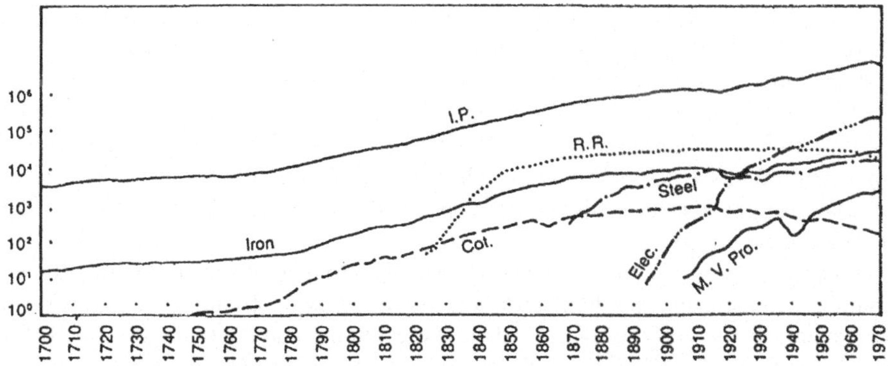

Fig. 5.4 Aggregate and Sectoral Growth Patterns Illustrated: British Industrial Production and Seven Major Sectors, 1700–1972. *Note* Industrial production (I.P.); railroad mileage (R.R.); raw cotton consumption (Cot.); production of iron (iron); steel (steel); electricity (Elec.); and motor vehicles (M.V. Pro.). *Source* Rostow, (1978) CHART III-2

Demand saturation is where the economics of Keynes and Schumpeter meet. Keynes argues for public investment and low interest rates to overcome recessions caused by insufficient demand, while Schumpeter argues that the key to overcoming low growth from demand saturation is innovation.

Saturation of demand is the fundamental factor restraining growth. However, in the standard literature, the fundamental factor restraining economic growth is diminishing returns to capital in production or R&D technology. In the less mathematical literature and casual discussions, the idea of 'demand saturation' has been very popular. Every businessman would acknowledge saturation of demand for an individual product. Figure 5.4, from Rostow (1978), demonstrates this 'stylized fact' on a historical scale. As we already saw, the experiences of diffusion of consumer durables such as refrigerators, television sets, cars, and personal computers tell us that deceleration of growth comes mainly from saturation of demand rather than diminishing returns in technology. Growth of production of a commodity or in an individual industry is bound to slow down because demand grows fast at the early stage but eventually, necessarily slows down. Thus, the demand for some products grows much more rapidly than the gross domestic product (GDP), while the demand for others grows much more slowly. Products/industries face different income elasticities of demand.

5.4 Product Innovations and Economic Growth

Unfortunately, the literature on growth abstracts largely from the important fact that products/industries obey the law of demand saturation and that each product/industry experiences a typical S-shaped life cycle. This, of course, is not to say that the appearance of new products and the disappearance of old ones have not been modeled.

The so-called "creative destruction" and the "quality ladder" literature, such as Grossman and Helpman (1991), Aghion and Howitt (1992), Caballero and Jaffe (1993), and Horowitz and Lai (1996), have analyzed such phenomena in growth models. However, in this line of research, the old products disappear only through the introduction of new products. Unless new products appear, the demand for the existing products remains the same. Therefore, it is possible for the economy to continue growing if it succeeds in raising productivity in the production of the existing commodities.[2]

With saturation of demand, however, raising productivity in the production of 'mature' products does not help sustain economic growth. To put it another way, in the existing R&D-based growth models, the economy can continue to grow if, for example, the automobile industry keeps raising the quality of cars, whereas in our model, it cannot because demand for cars saturates despite quality improvement.

Likewise, the product life cycle in the literature, such as Grossman and Helpman (1991), is based on a production technology or production geography life cycle, while the model explained in this section is based on a demand life cycle. In contrast to the "creative destruction" that occurs in the literature, the growth of demand for the existing commodities in our analysis of "saturation" necessarily slows down whether new commodities appear. It would be absurd to argue that the growth of demand for food decelerated, as Engel found because manufactured products appeared. The demand for cars did not approach its ceiling because personal computers were invented. Rather, the law of demand saturation works for an individual commodity.

Within the same industry, new and old products are often close substitutes, such as black/white and color TVs or personal computers of different vintages, and old products gradually disappear as new products appear. Thus, the "creative destruction" story fits the growth of an industry. The R&D race among competing firms, as modeled in the standard endogenous growth literature, certainly plays an important role. Technical progress taken up in the literature basically concerns close substitutes, as those models explicitly state. However, as we argued above, the same story does not necessarily hold true for different industries. Arguably, demand saturation is more relevant for the growth of the economy as a whole.

For a majority of economists, technical progress and innovations are supply-side problems. Schmookler (1966) using the U.S. patent data explored whether inventions are determined by (1) supply side factors, such as costs of R&D and progress of science and technology, or (2) the conditions of demand for goods and services that invention concerns. Perhaps contrary to many economists' prior expectations, his conclusion is that demand is the key factor that conditions the pace of inventions.

In what follows, we explore the growth model based on demand saturation and innovations that create new demand. To repeat, in the standard 'creative destruction',

[2] This point trivially applies to the case where the quality improvement occurs in the production of intermediate goods. It also applies to the case where the 'creative destruction' or the quality improvement occurs in the production of final consumables (see, for example, Eq. (21) of Caballero and Jaffe, 1993). If the number of commodities remains constant ($\dot{N} = 0$), the growth rate of the economy is $\hat{\eta}$, the growth rate of labor productivity in the production of the existing commodities.

'quality ladder' or 'product variety' models, the economy can sustain growth if productivity in the production of existing products keeps rising, while in our demand saturation model, it cannot.

We take the logistic growth of an individual product/industry as a 'stylized fact' and present a formal model of growth built on this 'stylized fact'. An obvious implication of the logistic growth of an individual product/industry is that the economy enjoys high growth if it successfully keeps introducing new products or industries that temporarily enjoy high growth of demand.

It is important to recognize that the demand-creating innovation is different from the standard total factor productivity (TFP) or an 'upward shift' of the production function. In the standard 'quality ladder' models and the "creative destruction" literature, such as Grossman and Helpman (1991), Caballero and Jaffe (1993), Horowitz and Lai (1996), and Young (1998), innovation or technical progress basically raises TFP by way of replacing old commodities with new ones simply because new commodities are assumed to have higher value than old ones. Again, whereas this may hold true for the commodities that are basically the same but of different vintages, the same story does not make much sense for wholly different products such as cars and personal computers. Personal computers do not necessarily command higher value added than cars. In short, the standard literature models the dynamics of close substitutes, while our model stresses the importance of demand saturation and the creation of wholly different products or industries for which demand grows transitionally.

The difference between the standard models and our model of demand saturation most clearly shows up in the transitory dynamics. In the standard R&D-based growth models, the efficiency of R&D determines the transitory dynamics, whereas in our model, the pattern of demand saturation is the determinant. As a model of demand-constrained growth, our model follows the long line of postKeynesian literature such as Kaldor (1957) and Robinson (1962). In the postKeynesian tradition, income distribution between capital and labor plays a central role in determining aggregate demand. In contrast, our analysis focuses on saturation of demand for an individual good/sector as a factor to generate demand constraints facing the economy.

Innovations in the economy facing the law of demand saturation contribute to growth in a different way than an 'upward shift' of the production function does. That TFP does not necessarily capture the significance of technological progress is pointed out by Wright (1997):

> The identification of 'technological progress' with changes in total-factor-productivity, or with the 'residual' in a growth-accounting framework, is so widely practised that many economists barely give it a passing thought, regarding the two as more-or-less synonymous and interchangeable...... Even with extensive quality adjustments, TFP is not generally a good index of technology. If a genuine change in technological potential occurs in a firm, an industry, a sector, or a country, in any plausible model this change will affect the mobilisation of capital and labour in whatever unit is involved. In the new equilibrium, inputs as well as outputs will have changed; the ratio between these may convey little if any useful information about the initiating change in technology (Wright, 1997, 1562).

We share Wright's concern. The economy always mobilizes resources and accumulates capital whenever it finds goods or sectors for which demand grows rapidly.

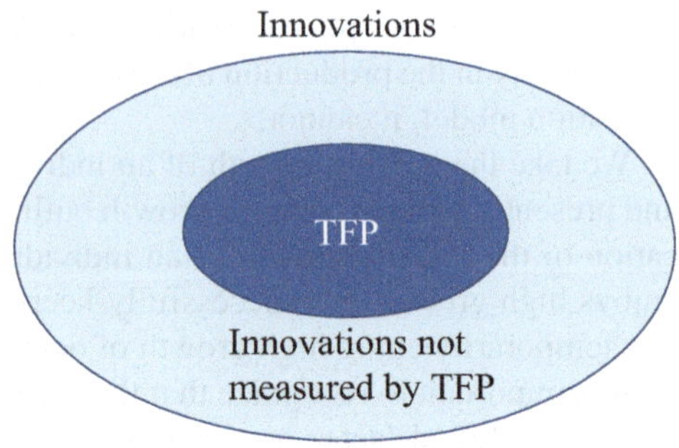

Fig. 5.5 Innovations and Total Factor Productivity. *Source* Aoki and Yoshikawa (2002)

In fact, in our model, the elasticity of capital in the production function is equal to one (the so-called AK model). Therefore, the economy grows whenever capital accumulates. However, capital accumulation is constrained by the saturation of demand. Innovation creates goods/sectors for which demand grows fast, elicits capital accumulation, and thereby ultimately sustains economic growth. Thus, demand-creating innovations are conceptually different from the growth of total factor productivity (TFP). The relationship between innovations and TFP is depicted in Fig. 5.5.

A Model of Demand-Creating Innovations

To substantiate this argument, we present a model that incorporates the basic idea that demand for an individual good or sector necessarily faces saturation and thus its growth eventually slows down (Aoki & Yoshikawa, 2002). Outside of the steady state, 'vigor of demand' and saturation determine growth, while the ultimate factor for sustaining economic growth in the steady state is the creation of new products/industries.

Output is determined by demand. That is, the output of each good is equal to its demand $D_k(t)$ no matter how the latter is determined.

$$y_k = D_k \tag{5.2}$$

Here, we take an S-shaped life cycle of demand for each product/industry as a stylized fact. To make our analysis tractable, assume that $D_k(t)$ follows the logistic curve:

$$D_t = \frac{\mu D_0}{[\delta D_0 + (\mu - \delta D_0)e^{-\mu t}]} \tag{5.3}$$

Since the mechanism is the same for all the products or sectors, for the moment, we drop k and write $D_k(t)$ as $D(t)$. D_0 in Eq. (5.3) is the initial value of D_0. Starting with D_0 smaller than μ/δ, $D(t)$ initially increases almost exponentially, but its growth eventually decelerates, and approaching its 'ceiling' μ/δ, the growth rate declines asymptotically to zero.

Although exponential growth is often taken for granted by economists, there is actually ample evidence to show that no individual product or industry grows exponentially. Rather, demand for or production of a product or an industry typically grows according to the logistic curve. In fact, an eminent physicist Montroll (1978) suggests that almost all social phenomena, except in their relatively brief abnormal times, obey logistic growth. Figure 5.3 demonstrates this well-known fact of life in our economy.

The growth and saturation of an individual product/sector are characterized by two parameters μ and δ. They would depend not only on preferences but also on the creation of new models and close substitutes of higher quality in the production of the same commodity. For example, TV set growth has certainly reached its ceiling much earlier if there had been only black/white TVs. The emergence of color TVs and small models pushed up the ceiling. We maintain, however, that such technical progress cannot overcome the law of demand saturation in the end.[3] In fact, based on his careful study of the U.S. patent data, Schmookler (1966) argues that technical progress in one industry is itself very strongly conditioned on the prospects of demand in that industry.

Following the logistic growth of demand, production $y_k(t)$ also satisfies (5.3). Thus far, we have focused on a final good. The number of final products is not given, however. Rather, at every moment, a new product or sector may arise. The emergence of an utterly new final good or a new sector is the result of innovations.

Much effort has been made to explicitly analyze R&D activities and inventions in growth models. In fact, the achievement of 'endogenous growth theory' is to have combined growth models with models of R&D activities. However, as we pointed out above, technical progress in the literature basically concerns quality improvement in the production of close substitutes. Therefore, it more closely applies to an industry than to the economy as a whole. We maintain that such technical progress "pushes up the ceiling" of demand for the existing products but cannot overcome the law of demand saturation in the end. Innovations in this model, in contrast, bear wholly new products/sectors/industries for which demand grows fast. Such innovations would depend not only on profit-motivated R&D but also on basic research. In any case, our primary interest is not in microeconomic foundations for R&D activities but in the way in which technical progress or innovation affects the macroeconomy.

We assume that an invention of a new final good or an emergence of a new sector stems stochastically from learning in the process of production of the existing products. Specifically, we assume that the probability that a new final good is invented or a new industry emerges between t and $t + \Delta t$ is $\lambda N \Delta t$, where N is the number of existing final goods ($\lambda > 0$). Since an invention or new sector is a branch off

[3] Kuznets (1953), for example, argues that: "in the industrialized countries of the world, the cumulative effect of technical progress in a number of important industries has brought about a situation where further progress of similar scope cannot be reasonably expected. The industries that have matured technologically account for a progressively increasing ratio of the total production of the economy. Their maturity does imply that economic effects of further improvements will necessarily be more limited than in the past". Based on the American experiences, McLaughlin and Watkins (1939) share this kind of pessimism.

from an existing good or sector, the rate of success probability is proportional to the number of existing final goods/sector N. The more products or sectors in the economy, the more likely a new product or sector emerges. Here, λ is a parameter that represents the strength of innovations or, more precisely, the probability that a new good or industry emerges in the existing process of production. Innovations are thus accidental but depend on the prior 'knowledge' and experiences that stem from the existing production.

Given this assumption, $Q(N, t)$, the probability that the number of final goods at time t, $N(t)$, is equal to N satisfies the following equation:

$$\frac{dQ}{dt} = -\lambda N Q(N, t) + \lambda(N - 1)Q(N - 1, t) \tag{5.4}$$

Without loss of generality, we can assume that the initial number of final goods is one:

$$Q(N, 0) = \delta(N - N_0) = \delta(N - 1)$$

The solution of this equation under this initial condition is

$$Q(N, t) = e^{-\lambda t}(1 - e^{-\lambda t})^{N-1} \tag{5.5}$$

The probability that there are N goods at time t and the $N + 1$-th good emerges between t and $t + \Delta t$ is then given by

$$\lambda N Q(N, t)\Delta t = \lambda N e^{-\lambda t}(1 - e^{-\lambda t})^{N-1}\Delta t \tag{5.6}$$

At time t, the production of the final good that emerged at $\tau(\tau < t)$, $y_\tau(t)$, has grown to

$$y_\tau(t) = \frac{\mu}{\delta + (\mu - \delta)e^{-\mu(t-\tau)}} \tag{5.7}$$

since the growth of $y_\tau\ (t)$ obeys the logistic curve. Again, without loss of generality, we can assume the initial production of newly invented good D_0 to be 1 in (5.3).

We now consider the growth of the economy as a whole. The aggregate value added or GDP of this economy is stochastic, but in what follows, we will focus on its expected value and denote it by $Y(t)$. $Y(t)$ is simply the sum of the production of all the final goods.

Figure 5.6 illustrates this model economy. Each sector once it emerged grows logistically. New sectors emerge stochastically, and the aggregate value added or GDP is simply the sum of outputs of all the then existing sectors.

From (5.6) and (5.7), we know that the expected value of GDP of this economy is given by

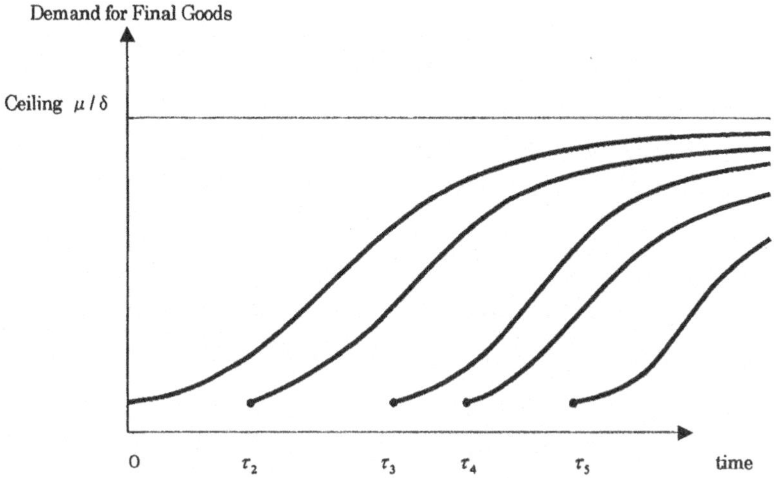

Fig. 5.6 Saturation of demand and emergence of new final goods or industries. *Note* τ_i is the date of birth of the ith good/industry. *Source* Aoki and Yoshikawa (2002)

$$
\begin{aligned}
Y(t) &= \sum_{N=1}^{\infty} \int_0^t \lambda N e^{-\lambda\tau}(1-e^{-\lambda\tau})^{N-1} y_\tau(t) d\tau + \frac{\mu}{(\delta + (\mu-\delta)e^{-\mu t})} \\
&= \sum_{N=1}^{\infty} \int_0^t \lambda N e^{-\lambda\tau}(1-e^{-\lambda\tau})^{N-1} \frac{\mu}{\left[\delta + (\mu-\delta)e^{-\mu(t-\tau)}\right]} d\tau \\
&\quad + \frac{\mu}{(\delta + (\mu-\delta)e^{-\mu t})}
\end{aligned}
\tag{5.8}
$$

The second term of the right-hand side is simply the output of the 'first' sector at time t, $y_0(t)$. Using

$$
\lambda N e^{-\lambda\tau}(1-e^{-\lambda\tau})^{N-1} = \frac{d}{d\tau}(1-e^{-\lambda\tau})^{N} \tag{5.9}
$$

and

$$
\sum_{N=1}^{\infty}(1-e^{-\lambda\tau})^{N} = e^{\lambda\tau} - 1 \tag{5.10}
$$

we obtain

$$
Y(t) = \int_0^t \frac{\left[\frac{d}{d\tau}(e^{\lambda\tau}-1)\right]\mu}{\left[\delta + (\mu-\delta)e^{-\mu(t-\tau)}\right]} d\tau + \frac{\mu}{(\delta + (\mu-\delta)e^{-\mu t})}
$$

$$= \lambda \int_0^t \frac{e^{\lambda\tau}\mu}{\left[\delta + (\mu - \delta)e^{-\mu(t-\tau)}\right]} d\tau + \frac{\mu}{(\delta + (\mu - \delta)e^{-\mu t})}$$

$$= \lambda \int_0^t \frac{e^{\lambda(t-u)}\mu}{\left[\delta + (\mu - \delta)e^{-\mu(t-u)}\right]} du + \frac{\mu}{(\delta + (\mu - \delta)e^{-\mu t})} \tag{5.11}$$

From (5.11), the growth rate of the expected value of GDP, g_t becomes

$$g_t = \frac{\dot{Y}(t)}{Y(t)} = \lambda + \left(\frac{f(t)}{Y(t)}\right)\left(\frac{\dot{f}(t)}{f(t)}\right) \tag{5.12}$$

where $f(t)$ is the logistic equation:

$$f(t) = \frac{\mu}{(\delta + (\mu - \delta)e^{-\mu t})}$$

It is easy to show that g_t satisfies

$$\dot{g}_t = (g_t - \lambda)\left[2(\mu - \delta)e^{-\mu t} f(t) - \mu - g_t\right] \tag{5.13}$$

with the initial value g_0.

$$g_0 = \left.\frac{\dot{Y}(t)}{Y(t)}\right|_{t=0} = \mu - \delta + \lambda \tag{5.14}$$

Additionally, since $e^{-\mu t} f(t)$ approaches zero, we can establish that the growth rate of GDP asymptotically approaches λ.

$$\lim_{t\to\infty} g_t = \lim_{t\to\infty} \frac{\dot{Y}(t)}{Y(t)} = \lambda \tag{5.15}$$

The growth rate of the economy is initially higher than λ by μ − δ, but it eventually goes down to λ. In this model, μ and δ have level effects, while the ultimate growth rate is determined by λ. The exact time path depends, of course, on all the parameters μ, δ, and λ.

It is important to recognize that not only steady state growth but also out-of-steady state growth is generated by the successive emergence of new products/industries. The growth of older industries continues to decline, while newer products/industries enjoy high growth. How high depends on μ and δ.

From the perspective of this model, it is easy to understand that historians have identified the 'leading' or 'key' industries in the process of economic growth. The best-known example would be perhaps Rostow, who argues that

> The most cursory examination of the growth patterns of different economies, viewed against a background of general historical information, reveals two simple facts. (1) Growth-rates in the various sectors of the economy differ widely over any given period of time. (2) In some meaningful sense, over-all growth appears to be based, at certain periods, on the direct and indirect consequence of extremely rapid growth in certain particular key sectors (Rostow, 1960, 261–262).

The vigor of the leading sectors depends on μ and δ in the model. For the sake of illustration, we show a simulation result (Table 5.1 and Fig. 5.7). In this example, we assume that λ, μ, and δ are 0.03, 0.12, and 0.02, respectively. Table 5.1 and Fig. 5.7 show both the growth rate of GDP and the average growth rate defined as $\sum_{\tau=1}^{t} g_\tau/t$ for each period (year). For the first 10 years, the growth rate of the economy is higher than 9 percent. In year 20, it is still 5.7 percent. By year 40, the growth rate slowed to 3.2 percent, which is close to the assumed asymptotic rate of 3 percent. The average growth rate, of course, decelerates much more slowly than the growth rate. The average growth rate for the first 30 years, for example, is 7.5 percent, although the growth rate in year 30 is 3.9 percent. This example demonstrates that depending on μ and δ, the economy can sustain a much higher growth rate than the equilibrium rate for a very long period. The transitionary dynamics are determined by the pattern of demand saturation.

Everyone knows that no economy grows at 10 percent indefinitely. Some economies, however, actually experienced 10 percent growth for a decade, and this decade-long high growth is often crucial for their growth experiences. Japan, for example, experienced 10 percent growth for a decade and a half, from 1955 to 1970. As of 1955, almost half of the working population in Japan was in agriculture. The era of high economic growth transformed a semitraditional economy into a modern industrial nation. After several decades, China also experienced 10 percent growth during the periods of 1992–96 and 2003–11 and became the second largest economy in the world. Certainly, we cannot dismiss the 'out-of-steady state' merely as transitory but must attach equal importance to it as to the steady state.

Table 5.1 Simulation results

Time	GDP	Growth rate (%)	Average growth rate (%)
0	1.00		
1	1.14	12.8	12.8
3	1.45	11.8	12.3
5	1.81	11.0	11.9
10	2.95	9.0	10.8
20	6.01	5.7	9.0
30	9.48	3.9	7.5
50	18.26	3.1	5.8
70	33.37	3.0	5.0
100	82.09	3.0	4.4

Source Aoki and Yoshikawa (2002)

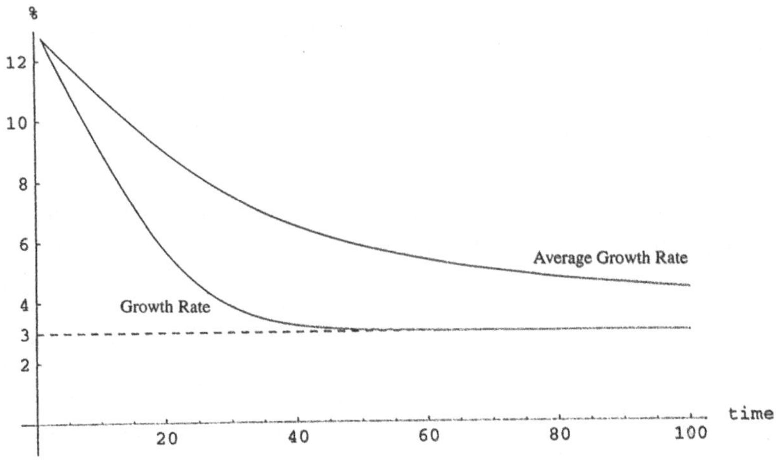

Fig. 5.7 Simulation results. *Source* Aoki and Yoshikawa (2007)

The out-of-steady state growth path illustrated in Fig. 5.7 is qualitatively similar to that obtained in the old Solow (1956) model and in the more recent R&D-based growth models such as Jones (1995), Young (1998) and Segerstrom (1998). Namely, the growth rate decelerates over time. The mechanism is fundamentally different, however. In the Solow model, diminishing returns to capital in ordinary production is the factor that brings about slower growth. In the R&D-based models, diminishing returns in R&D lead to slower growth. In contrast, in the present model, the deceleration of the out-of-steady state growth rate comes from the saturation of demand. To be specific, as (5.13) shows, the out-of-steady state growth path depends on μ and δ, which determine how vigorous growth of demand is, or conversely, how soon demand reaches its saturation. When δ is very small, and therefore, the 'ceiling' of demand is very high, the pace of deceleration can be slow. This out-of-steady state growth is basically constrained by demand. In this sense, the present model follows the long tradition of postKeynesian literature mentioned earlier.

The Ramsey model is the most standard microfoundation. However, in many economies, for many periods in history, a declining growth of demand for a particular product has been very closely related to the diffusion of the product among *different* households. Some households own the product, while others do not. This is particularly true for consumer durables such as televisions, refrigerators, cars, and personal computers. In Sect. 5.2, we have seen how the diffusion of consumer durables brought the rapid growth of the Japanese economy during the 1950s–1960s to end. Bass (1969), referred to earlier, also underlines this fact using the U.S. data. Thus, it makes much more sense to analyze growth in a model with different households than in a model with the representative consumer. In what follows, we consider such a model.

Suppose that there are M households in the economy. Without loss of generality, we can assume that M is equal to μ/δ. Households are indexed by i ($i = 1, 2..., M = \mu/\delta$). We define $f_{iN}(t)$ functions:

$f_{iN}(t) = 1$, if household i purchases the Nth product at time t;
$f_{iN}(t) = 0$ if household i does not purchase the Nth product at time t.

We assume that the number of households that consume the Nth product at time t $m_N(t)$ follows a birth and death process with birth rate μ and death rate δm_N. Note the following relation:

$$\sum_{i=1}^{M} f_{iN}(t) = m_N(t) \tag{5.16}$$

This process leads us to the logistic equation for the expected value of $m_N(t)$, $\widehat{m}_N(t)$. Thus, if the Nth product emerged at τ, $\widehat{m}_N(t)$ satisfies the following equation:

$$\widehat{m}_N(t) = \frac{\mu}{\left(\delta + (\mu - \delta)e^{-\mu(t-\tau)}\right)} \quad \text{for each } N \tag{5.17}$$

The (expected) diffusion rate or the percentage of households that consume the Nth product is $\widehat{m}_N(t)/M$.

For simplicity, we assume that household purchases $1 - s$ unit of any final product if it consumes this product; s is the saving rate. As in Solow (1956), the present analysis abstracts from the determination of s. The saving rate is assumed to be common for all the households ($i = 1, 2, \ldots, M$), and depends possibly on the interest rate. It also depends on time t. Note that the consumption of a final good by each household is constant. This assumption seems to hold, as an approximation, for many consumer durables.

Then, with the expected income of household i, $I_i(t)$, the budget constraint for household i at time t becomes

$$\begin{aligned} I_i(t) &= \sum_{N=1}^{\infty} \int_0^t \lambda N e^{-\lambda\tau}\left(1 - e^{-\lambda\tau}\right)^{N-1}(1-s) f_{iN+1}(\tau)d\tau + (1-s) f_{i1}(t) + sI_i(t) \\ &= \sum_{N=1}^{\infty} \int_0^t \lambda N e^{-\lambda\tau}\left(1 - e^{-\lambda\tau}\right)^{N-1} f_{iN+1}(\tau)d\tau + f_{i1}(t) \end{aligned} \tag{5.18}$$

Thanks to (5.16) and (5.17), the incomes of all the households (5.18) sum up to GDP, $Y(t)$:

$$\sum_{i=1}^{M} I_i(t) = \sum_{N=1}^{\infty} \int_0^t \lambda N e^{-\lambda\tau}\left(1 - e^{-\lambda\tau}\right)^{N-1} \sum_{i=1}^{M} f_{iN+1}(\tau)d\tau + \sum_{i=1}^{M} f_{i1}(t)$$

$$
\begin{aligned}
&= \sum_{N=1}^{\infty} \int_0^t \lambda N e^{-\lambda\tau}\left(1 - e^{-\lambda\tau}\right)^{N-1} \frac{\mu}{\left(\delta + (\mu - \delta)e^{-\mu(t-\tau)}\right)} d\tau \\
&\quad + \frac{\mu}{(\delta + (\mu - \delta)e^{-\mu t})} \\
&= Y(t) \qquad (5.19)
\end{aligned}
$$

The S-shaped growth derives from the diffusion of each product among different households. The growth of demand induces capital accumulation by the demand-constrained firm and thereby economic growth. Growth, on the other hand, creates higher income, which makes more households able to purchase goods. Equation (5.18) defines income distribution, which generates the diffusion of final goods among households. Because the amount of a final good that each household purchases is bounded, the growth of the production of an individual good necessarily decelerates parallel to the diffusion of the good among households. The creation of wholly new goods/sectors is the ultimate factor in sustaining growth in such an economy.

Implications of the Model

In the standard literature, the fundamental factor restraining economic growth is diminishing returns to capital in production or R&D technology. In the present model, the factor restraining growth is the saturation of demand. This model provides new perspectives on several important problems addressed by the economics of growth.

The first new perspective pertains to out-of-steady state dynamics. Despite the controversy surrounding conditional convergence, for many economies, and for advanced economies in particular, we do observe the eventual deceleration of growth rates. In the Solow model, diminishing returns to capital in production are the factor bringing about slower growth. R&D-based models have diminishing returns in R&D. Thus, in both approaches, diminishing returns in technology are the factor that brings about a slowdown of economic growth. In contrast, in our model, saturation of demand (μ and δ) is the factor that leads the economy to slower growth.

The second new perspective relates to the nature of technical progress or innovations. In the standard analysis, technical progress brings about higher value added given the same level of inputs. It is basically equivalent to an 'upward shift' of the production function. The so-called 'product variety', 'quality ladder' or 'creative destruction' models, such as Grossman and Helpman (1991) and Caballero and Jaffe (1993), using the Dixit/Stiglitz production/utility function, successfully endogenize this kind of technical progress. Empirically, technical progress has been measured by growth accounting as TFP.

In the model presented here, the aggregate production function is $Y = AK$. Since A is constant, there is no TFP growth. For the economy to grow, capital must accumulate. However, saturation of demand constrains capital accumulation and leads the economy to deceleration of growth. Innovation or technical progress in this model creates a major new product or industry that commands high growth of demand and thereby elicits capital accumulation and ultimately sustains economic growth.

As stated above, the concept of innovation is broader than TFP (Fig. 5.5). Schumpeter (1934), in his famous book, distinguishes five types of innovations: (1) the introduction of a new good; (2) the introduction of a new production method; (3) the opening of a new market; (4) the conquest of a new source of supply of raw materials; and (5) the new organization of industry. His first and third types of innovations as an engine for growth are most naturally interpreted in terms of the kind of model presented here.

The distinction between the conventional TFP and demand-creating technical progress is not only theoretically important but is also empirically relevant. Young (1995), for example, in his careful study of growth accounting, demonstrates that TFP growth in the then newly industrializing countries (NICs) of east Asia, i.e., Hong Kong, Singapore, South Korea, and Taiwan, is not so extraordinarily high but actually comparable to those in other countries. The very high average growth rate (8–10 percent) of East Asian NICs for such a long period as 25 years must therefore be explained by the extraordinary injection of capital and labor rather than extraordinary TFP growth.

Nevertheless, the basic question remains: why did the economy grow so fast in these countries? The present analysis suggests that in these countries, new sectors that command high growth of demand vigorously emerged (high μ/δ and λ). TFP growth may not be extraordinarily high, but it does not necessarily mean there is an absence of demand-creating innovations. Young, in fact, reports that in East Asian NICs, the industrial structure has drastically changed. A change in the industrial structure in the course of economic growth must reflect demand-creating innovations that are conceptually different from TFP. To the extent that exports happen to have commanded high growth (i.e., high μ/δ), we can easily understand that the high growth of East Asian NICs often looked to be export-led. China's high growth was also led by exports.

To repeat, the present model shows that technical progress has a different aspect than the conventional TFP represents, namely, demand creation, and that it is demand creation that sustains economic growth. It sheds light not only on 'the Asian/Chinese miracle' but also on the high growth of the American economy during the 1990s, where despite remarkable progress in information technology (IT), TFP growth was not as high as one might expect at least until the late 1990s. Advancement of information technology may have created new markets and mediate demand-led growth if its contribution to productivity on the supply side is limited.

The growth of TFP has decelerated all over the world since the 1980s. Gordon (2016) argues that great innovations have already occurred and that only those of lower quality remain. Bloom et al. (2020) in the paper entitled "Are Ideas Getting Harder to Find?" underlines a secular decrease in the productivity of research. The simple fact that over seventy years from 1930 to 2000, the U.S. TFP growth continued to decline, whereas the number of researches grew increasingly quickly during the same period (Fig. 1 in their paper).

However, research productivity with input measured by the number of researchers cannot fully capture the strength/weakness of demand-creating innovations. A case in point would be the COVID-19 upheaval. Many traditional shops were forced to close down during social confinement. At the same time, new online services have emerged. This process of creative destruction, which would have sustained growth with little change in TFP, largely reflects changes in demand and has little to do with the number of researchers.

The ultimate factor in sustaining growth, λ, which creates wholly new products/industries, is not necessarily confined to supply side factors. An appropriate income distribution policy that triggers the diffusion of major products can be taken as a demand-creating innovation. Conversely, inequality may hinder economic growth. More broadly, the integration of Keynes's principle of effective demand for the short run and growth theory for the long run remains a central theme in macroeconomics. The model presented here may provide a constructive step toward solving this problem.

5.5 Population and Economic Growth

The matter of population has drawn people's interest since ancient times. As eighteenth-century Europe was undergoing a population explosion, it was only natural that the economists, who were pursuing an emerging academic discipline, discussed population amid the great changes in the economy caused by the Industrial Revolution. The standard way of thinking at the time was that a large population symbolizes social affluence, while conversely, a small population means poverty. In his 1776 book *Wealth of Nations*, Adam Smith (1723–90) wrote "The most decisive mark of the prosperity of any country is the increase of the number of its inhabitants" (book I, chapter-I/114 "On the Wages of Labor"). Note that Smith does not say "the number" of inhabitants, but rather "the increase of the number." Here, Smith also argues that it is not the size of GDP (gross domestic product) but rather its continual increase, namely, *growth* that causes a rise in (real) wages.

Smith was not the only one who believed that an increasing population was a positive factor for the economy. Schumpeter, in his *History of Economic Analysis*, wrote that through the middle of the eighteenth century, all economists were proponents of population growth.

> With rare exceptions, they were enthusiastic about "populousness" and rapid increase in numbers. In fact, until the middle of the eighteenth century, they were as nearly unanimous in this "populationist" attitude as they have ever been in anything. A numerous and increasing population was the most important symptom of wealth; it was the chief cause of wealth; it was wealth itself—the greatest asset for any nation to have (Schumpeter, 1954, 251).

It was Robert Malthus who then caused a sensation in the population debate. There is no book that has had a greater impact on future generations than *An Essay on the Principle of Population* (first edition, 1798) by Malthus, who lived during the period of great population growth in Europe. He postulated that whenever the level of income rises, the population necessarily increases. A century later, his postulate turns out to be wrong.

Brentano (1910) investigated the demographic trends in Europe at the end of the nineteenth century and discovered that birthrates were *declining* even though average income levels were *rising* in every country. Based on this, he raised objections to Malthus' thesis. According to Malthus, when income levels rise, people should have many children, and the population should grow, but Brentano discovered that the opposite was occurring. It is true that a country's overall birth rate is influenced by marriage rates, but he found that in fact, it was a decline in the number of children per married couple that had a large effect on the decline in the birth rate. Moreover, through a detailed examination of the birthrate in each country of Europe over approximately thirty years from the end of the nineteenth century through the beginning of the twentieth century, looking separately by region, income level, and occupation, Brentano found that while the birthrate was not declining in households with low incomes, it was falling noticeably among those with high incomes and educational levels. Brentano called this a "new reality" that cannot be explained by the principle of Malthus.

Why did the birthrate decline among wealthy people? Brentano addressed several points of debate from the prior research that are still being examined by experts today. Along with the progress of society, the variety of goods and services enjoyed by young people expands, and enjoying them takes more time and money. As a result, people avoid childbirth and childbearing, which require time and money. People have fewer children, so they can maintain their high standards of living. They also want to give a small number of children access to higher education and allow them to enter professional occupations. In addition to such trends, changes in women's awareness have also been noted.

As one can easily imagine, it was wealthy people who were first affected by the changing times in which birthrates fell due to new entertainments such as going to the opera and traveling abroad, as well as changes in women's awareness and high educational levels. While the traditional birthrate was maintained among the middle- and lower-income segments, it was initially in the wealthy segment where the birthrate fell. When civilization prospers, contrary to what Malthus said, the population declines. Brentano's paper concludes by pointing out that the Roman Empire also declined in this manner.

To be sure, looking back over the long history of humanity, late nineteenth century Europe was not the first place where a wealthy nation experienced population decline. For example, Polybius, who was alive in the middle of the second century B.C. wrote about Greece at that time.

> In our time all Greece was visited by a dearth of children and generally a decline in population, owing to which the cities were denuded of inhabitants, and a failure of productiveness resulted, though there were no long-continued wars or serious pestilences among us... For this evil grew upon us rapidly, and without attracting attention, by our men becoming perverted to a passion for show and money and the pleasures of an idle life, and accordingly either not marrying at all, or, if they did marry, refusing to rear the children that were born, or at most one or two out of a great number, for the sake of leaving them well off or bringing them up in extravagant luxury.
>
> (Polybius, The Histories, BookXX. V111:9)

History repeats itself. The population started declining last century in most advanced countries. A century after Malthus' death, in 1937, Keynes gave Galton Lecture at a Malthus-related population research center. This was one year after Keynes published his main work, *General Theory* (1936). The lecture was entitled "Some Economic Consequences of a Declining Population."

In contrast to Malthus' era, when he never stopped expounding the evils of population growth, the problem facing Britain in Keynes's heyday during the 1920s and 1930s was the impact on the economy from population decline. This was not limited to Britain. In France, which had been defeated in the Franco-Prussian War (1870–71), there were already warnings of a crisis at the end of the nineteenth century, as France's population began to decline in comparison to that of its neighbor Germany, which was growing. The times had changed: population decline had already become a major problem.

Keynes pointed out that England had reached a major turning point, going from an era of population growth to an era of population decline; he said that the future would be a completely different world from the past. Above all, Keynes emphasized the influence of population on investment. Unlike the situation when Malthus and Ricardo were writing, manufacturing had already become the leading industry. Manufacturing requires large-scale capital stock, such as factories and machines that are in an entirely different dimension from the plows and hoes used for agriculture. What increases capital is the capital investment each year.

Gross domestic product, or GDP, consists of consumption, private investments, exports, and other government expenditures. Unless something extraordinary happens, consumption does not change drastically; the major exception is COVID-19, which caused a fall in private consumption during 2020–21. In contrast, corporate investment is an "unruly horse." The increase and decrease in this investment generate the business cycles that accompany a capitalist economy: This is what Keynes argued in his *General Theory*. When considering the long-term change from a growing population to a shrinking population, Keynes first examined the effect on investment.

What factors determine investment over the long term? Population is certainly one, as it determines the scale of the economy. Nevertheless, it is, of course, not the only factor. Technological advances, which boost our standard of living, also have a large effect on investment. The durability of capital is another important factor that determines investment.

According to Keynes, when we look over history from that perspective, large and highly durable objects were characteristic of Victorian civilization in the nineteenth century. The civilization of the twentieth century, compared to the Victorian era, is an era of lighter capital goods. This point made by Keynes is naturally convincing today, when highly convenient goods are best characterized as being small and lightweight.

Capital stock that is highly durable and large naturally requires a sizable investment. However, a small investment is sufficient for small, lightweight capital. If we can no longer expect highly durable, large capital stock, then population growth and technological advances that raise the standard of living are the only factors remaining to generate investment. However, we cannot expect enough technological advances to boost investment sufficiently. Moreover, the population is declining. With this reasoning, Keynes warned that, if left unchecked, the growth of investment in Britain after World War I would inevitably lack vigor.

If investment, which may be called the engine of capitalism, is insufficient, the economy falls into a recession. This is the conclusion of the *General Theory* published in 1936. There may be people who want to work, along with sufficient machinery and factories to produce goods, yet plants and equipment stand idle, and unemployment occurs. This is because the goods cannot be sold even if they are produced: the demand is insufficient. Such a recession may be characterized as "poverty in the midst of plenty."

The academic forerunner to the "macroeconomics of insufficient demand"' stressed by Keynes was Malthus. Contrary to Ricardo, Malthus defended the land-owning class. In response to criticism that the extravagant consumption by the land-owning class was simply wasteful spending, such as hunting, which gives no benefit to society, Malthus stated that what seems at first glance to be wasteful consumption by the land-owning class generates employment.

Keynes named the vice and misery brought forth by excessive population, as explained by the young Malthus, "the Devil P" (for "population"). On the other hand, he called the unemployment problem that Malthus noted when he was older "the Devil U" (for "unemployment"). The Devil P had made itself felt throughout the long history of mankind through to the first half of the nineteenth century. However, in the twentieth century, with a declining population, Britain was exposed to the Devil U as a result of insufficient investment.

Under a declining population, what is required to keep Devil U at bay and enjoy abundance? Keynes argued that consumption must increase to compensate for insufficient investment. In the nineteenth century, when there were any number of investment opportunities, savings were linked directly to investment, and therefore savings by the rich contributed to the progress of the economic system through capital accumulation. However, that era ended. Keynes argued that because robust investment cannot be expected in the twentieth century, consumption has to uphold effective demand in place of investment. To those ends, there must be income redistribution from the wealthy who save their income to the general public who consume.

Keynes's concern permeated Alvin Hansen's 1938 Presidential Address to the American Economic Association (Hansen, 1939). In this address, Hansen advanced the famous doctrine of *secular stagnation* for a mature economy such as the U.S. Like Keynes, he argued that a declining rate of population growth in most industrialized economies necessarily weakens the need for "widening of capital" to keep the same capital/labor ratio and thereby discourages investment. Thus, he continues, to avoid a large increase of unemployment in a mature economy, a substantial fiscal stimulus and an aggressive monetary policy to lower interest rates are imperative.

Samuelson (1988) formally demonstrates that a declining rate of population growth and the ensuing decline of the "natural growth rate" indeed raise the level of insufficient demand and the unemployment rate in a simple textbook multiplier-accelerator model. The point is, as Keynes and Hansen predicted, that investment

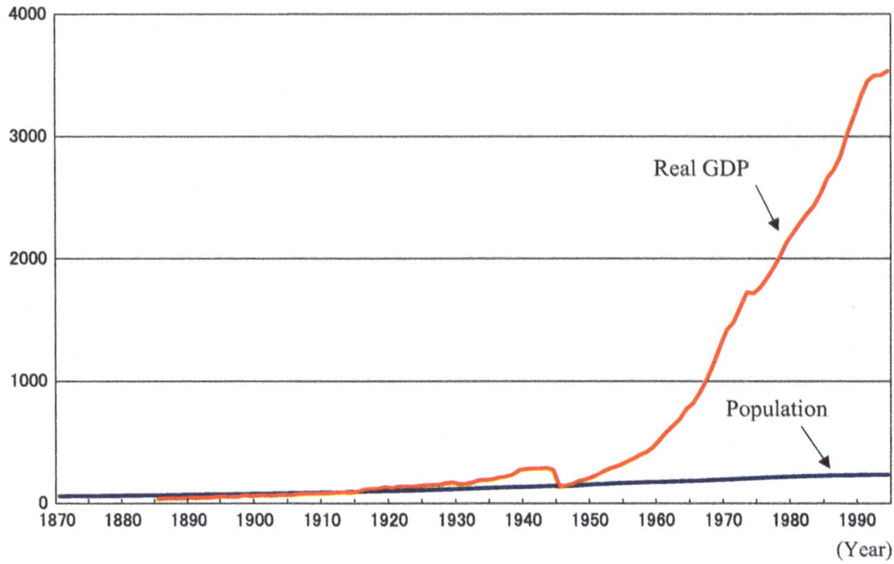

Fig. 5.8 Japan's population and economic growth (1870–1994). *Note* Population and real GDP both indexed with 1913 = 100. *Source* Maddison (1995)

is discouraged by a declining rate of population growth. However, according to the life cycle theory of consumption, savings in the economy as a whole are also discouraged by a declining rate of population growth (Modigliani & Brumberg, 1954; Modigliani & Ando, 1963; Modigliani, 1986). Samuelson demonstrates that although the life cycle effects quantitively reduce the stagnation effects of population decline, they do not negate the qualitative intuitions of Keynes and Hansen.

Now, in the twenty-first century, population decline prevails globally. Even China is expecting its population to start declining very soon. The number of births in 2021 was only 10.6 million against the current population of 1.4 billion, the lowest figure since 1949 when the People's Republic of China was established.

A declining population is in and of itself a negative factor for economic growth. However, quantitatively, the growth of *per capita* income dominates the impact of demography in advanced countries. Figure 5.8 shows the real GDP and population of Japan over a hundred and twenty years. A glance of the figure reveals that the growth of per capita GDP, which is generated by innovations, far dominates demography. Thus, the basic problem still appears to be innovations rather than population (Yoshikawa, 2020).

Chapter 6
Prices, Wages, and Monetary Policy

This book concerns macroeconomic theory. However, mainstream microfounded macroeconomics is not confined to academia but has significant impacts on real-world economic policy. A prime example is monetary policy under zero interest rates. The controversy concerns inflation/deflation.

Inflation together with unemployment is one of the most important issues driving policy decisions. More recently, deflation has been regarded as a threat to the macroeconomy. Bernanke (2020, 978) concludes his AEA Presidential Lecture, saying that "the Federal Reserve and other central banks should defend against inflation that is too low at least as vigorously as they resist inflation that is too high."

In the late 1990s, Japan was the first of the advanced countries to become trapped in deflation when the country faced the zero interest bound. The efficacy of monetary policy was seriously questioned. Krugman (1998) advanced the use of quantitative easing (QE) instead of the traditional interest policy coupled with a firmly committed inflation target as a possible remedy. Many others have elaborated on the idea. Today, most central banks are indeed committed to an explicit inflation target, such as an annual two-percent increase in the consumer price index (CPI). Facing the zero interest bound, they struggle against deflation by resorting to *unconventional monetary policies* such as quantitative easing, forward guidance, and negative interest rates. In passing, Bernanke (2020) argues that so-called "unconventional" policies should become part of the standard toolkit and, therefore, should be better referred to as "new" or "alternative" monetary tools. Regardless, the efficacy of these policies with respect to price depends, of course, on how prices are determined. In this chapter, we consider the determination of nominal prices and wages in relation to monetary policy.

According to microfounded macroeconomics, expectations play a crucial role in unconventional monetary policy such as quantitative easing (QE), which is said to be effective even under zero interest rates. On 27 August 2016, Haruhiko Kuroda, the Bank of Japan governor, gave a speech in Jackson Hole, USA. The title of his speech was "The reanchoring of the expected rate of price inflation through 'quantitative

H. Yoshikawa, *Reconstruction of Macroeconomics: Methods of Statistical Physics, and Keynes' Principle of Effective Demand*, Advances in Japanese Business and Economics 3, https://doi.org/10.1007/978-981-19-5264-7_6

and qualitative monetary easing with negative interest rates'". Bernanke (2020, 978) also argues that "And, of course, in any case, the Fed must work to keep inflation expectations from falling below target." The key to the effectiveness of monetary policy is supposed to be "expected rate of inflation," or "expectations" about price increases.

Needless to say, expectations have major impacts on stock prices, exchange rates and other asset prices. We know very well that the expectation that VAT will be raised in the near future creates more or less a "rush" for consumption. Expectations play a major role in economic activity in such cases. The question we are interested in this chapter is whether the proposition that "an increase in the quantity of money (money stock), even under zero interest rates, creates an expectation that prices will rise in the future, and thus prices will actually rise" is correct.

6.1 Deflation and the Economy

Before considering prices and wages, to provide some background, we briefly look back at deflation and the Japanese economy. Japan is a prime example of the economy that lapsed into deflation after the war; even the term "Japanization" was coined. For example, in 2015, as the financial crisis in Greece deepened, there were suggestions that the EU economy was about to be "Japanized".

Why did the Japanese economy fall into prolonged stagnation? What were the causes? What should have been done to get out of the prolonged stagnation? These are the most important questions to ask not only for the Japanese economy but also for the global economy. The idea that the main cause of the long-term stagnation is a "sustained decline in the general price level," or deflation, has been a prevailing theory among prominent macro economists since the end of the 1990s and was, in fact, the central idea behind the economic policy under the administration of Mr. Shinzo Abe's cabinet or *Abenomics,* which took off in December 2012. Hamada (2013), an official advisor to the Abe administration, is representative of this position.

> Since the new Bank of Japan Act was enacted in 1998, Japan's economy has had one of the worst macroeconomic performances of any country in the world, as will be shown in the next chapter. The main reason is that the Bank of Japan's monetary policy has continued to implement austerity measures over the past 15 years that have resulted in deflation and an extremely strong yen.
>
> Today, people suffer greatly because of deflation and the strong yen. Deflation concerns the price of the yen relative to goods, and a strong yen is the price relative to a foreign currency —— in other words, it is a monetary problem. Therefore, it can be resolved exclusively through monetary policy, and it is the Bank's responsibility to deal with it by monetary policy. (Hamada, 2013, 25–27)

Is deflation the primary cause of Japan's prolonged economic stagnation? Until March 2013, when Mr. Masaaki Shirakawa stepped down as governor of the Bank of Japan, the Bank of Japan had repeatedly stated that deflation was the *result* of stagnation in the real economy. Conversely, the basic idea behind Abenomics and the "Kuroda Easing" is that deflation is not a consequence of economic stagnation but is the main *cause* of the problem.

After the bursting of the bubble economy at the beginning of the 1990s, asset price deflation —— land and stock prices —— generated the huge amount of nonperforming loans, and the Japanese economy entered a long stagnation. There is a general consensus among Japanese economists that financial troubles that culminated in the 1997-98 financial crisis made a major impact and eventually let the Japanese economy fall into the "lost decade".[1] The vicious circle of debt and deflation that Fisher (1933) highlighted as the greatest threat to capitalist economies is the intertwining of bad debts and deflation, and it is exactly what happened to asset price deflation in Japan in the 1990s.

On the other hand, the present issue is whether the deflation of prices of goods and services, the subject of this chapter, is the main cause of the "lost 20 years", that is, whether the fundamental problem facing the Japanese economy is monetary. As we can observe in Fig. 6.1, Japan lapsed into mild deflation during the 2000s. However, we believe that the fundamental problems of the Japanese economy since the beginning of the 2000s, especially after the injection of public funds into *Risona Bank* in 2003, which forced an end to the decade-long bad loan problem, had been "real" and that mild deflation was not the *cause* of Japan's economic stagnation. Since 2003, Japan's economy has not had the bad loan problem that Fisher and Keynes faced in the 1930s.

Incidentally, the first half of the 19th century was the period when the British economy was at its most prosperous, and as a result, Britain became the "British Empire". In terms of real economic growth, the British economy was at its peak during this period. Specifically, between 1815 and 1847, industrial production continued to increase at an average annual rate of 3.5%. However, this period of prosperity was also a period of chronic deflation similar to that of Japan. Namely, prices continued to fall at an average rate of −1.4% for more than thirty years (Rostow, 1948, Table 1). This is a major example demonstrating that deflation, particularly mild deflation, does not always produce economic stagnation.

Borio et al. (2015) formally questions the thesis that deflation is very costly and tests the historical link between output growth and deflation in a sample covering 140 years up to 38 economies. They found that this link is weak and derives largely from the Great Depression. They find a stronger link between output growth and asset price deflation, particularly postwar property price deflation. Their finding is in good accordance with Japan's experience.

In Japan, the slump in nominal GDP is often cited as a symbol of the long-term stagnation of the Japanese economy. Fig. 6.2 shows the real GDP and nominal GDP for the past 26 years from 1994 to 2020. Real GDP increased from 446.5 trillion

---

[1] Hayashi and Prescott (2002) do not agree with this view, however. Based on RBC, they argue that financial problems were a minor factor in Japan's economic stagnation in the 1990s. When the U.S. economy plunged into a financial crisis after the collapse of Lehman Brothers in September, 2008, RBC seems to have finally faded away.

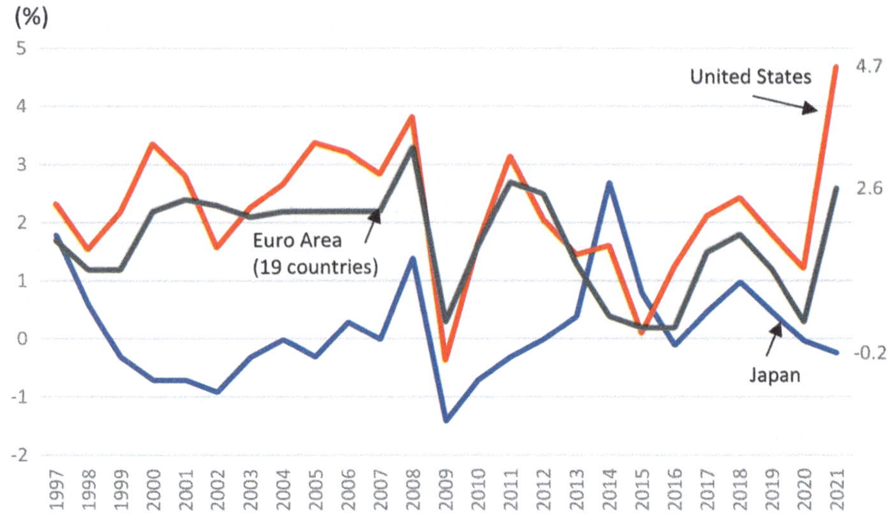

Fig. 6.1 Consumer price index (% changes year over year): International comparison. *Source* OECD (2022), Inflation (CPI) (indicator)

yen in 1994 to 528.2 trillion yen in 2020 despite the large decline during the 2008–09 Great Recession. In contrast to the growth in real GDP, nominal GDP *fell* by 8.5% for 14 years from 1997 to 497.4 trillion yen in 2011. It is often suggested that this symbolizes Japan's deflation. However, the matter is not so simple because the decline in the GDP deflator is not 100% monetary but reflects a "real" factor of worsening *terms of trade*. The terms of trade represent the number of units of import goods that one unit of export goods exchanges for. Hence, it is the "relative price" of import and export goods and is "real", not monetary. In a closed economy without imports and exports, the GDP deflator is a purely nominal price index, but when imports and exports are present, it is affected by the real terms of trade. Thus, in an open economy, the GDP deflator is actually a hybrid of nominal and real.

In fact, since the beginning of the 2000s, the price of primary commodities, which are important imports for Japan, has risen sharply. For example, the price of crude oil, which was 20-30 dollars a barrel in the early 2000s, rose to over 120 dollars in a decade. Figure 6.3 shows the terms of trade of Japan and Germany and crude oil prices. Like Japan, Germany is a resource-importing country, but its terms of trade deteriorated by only approximately 5% between 1995 and 2012. In contrast, Japan's terms of trade deteriorated by 40 percent over the same period. Germany could pass on a considerable increase in the price of imported raw materials to the price of its exports, whereas Japanese firms failed to raise the prices of their exports. This created a marked difference in the terms of trade of the two countries. The "trade losses" caused by the worsened terms of trade were 21 trillion yen (3.8% of GDP) in 2007, 23 trillion yen (4.3% of GDP) in 2008, and continued to 16.4 trillion yen (3% of GDP) in 2015. This worsening of the terms of trade resulted in a decline in the GDP deflator,

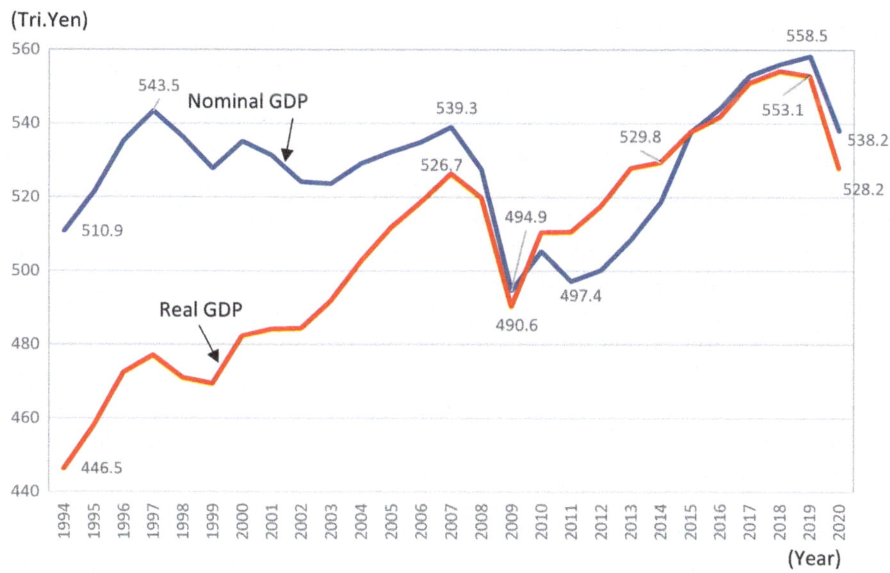

Fig. 6.2 Japan's GDP (1994–020). *Source* Cabinet Office, *SNA (National Accounts of Japan)*

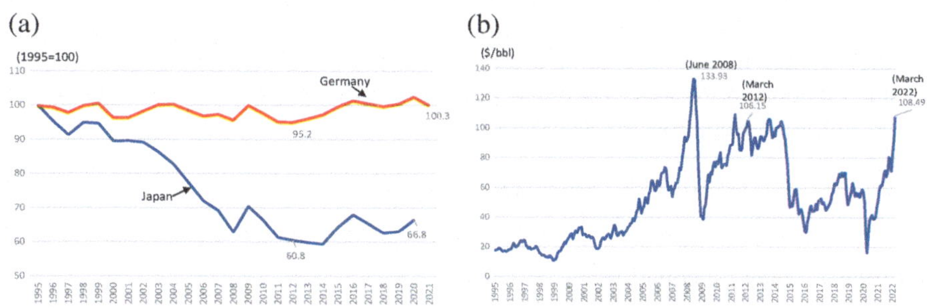

Fig. 6.3 (a) Terms of Trade of Japan and Germany. (b) Crude Oil Price (WTI). *Source* (a) OECD (2022), Terms of trade (indicator). (b) World Bank, Commodity Markets

or "deflation". As mentioned earlier, this is not a monetary phenomenon supposedly related to money stock but deflation created by the real factor of worsening terms of trade.

6.2 Monetary Policy and Expectations

Whether deflation is the most serious problem for the economy is open to debate. Here, we will not go further into this problem but rather consider the question of whether to increase money stock or quantitative easing (QE), arguably the most important tool of unconventional monetary policy, is effective in achieving the central bank's target of a 2 percent CPI increase, particularly in relation to expectations.

Monetary policy under normal economic conditions is an interest rate policy; for Japan, see Yoshikawa (1993). Monetary tightening means an increase in the interest rate, while monetary easing is a reduction in the interest rate. Since interest rate is a variable that has a significant impact on capital investment by firms and/or housing investment by households, there is no mystery about the transmission mechanism of monetary policy if how effective remains uncertain, namely, the problem concerning the interest elasticity of investment. Indeed, in a deep recession, the economy may lapse into an "uncertainty trap" in which interest elasticity becomes very small (Aoki and Yoshikawa, 2007, Chap. 4). In such a case, the effectiveness of monetary policy, either conventional or unconventional, must be necessarily weakened. The present problem is different, however. Beginning with Japan in the 1990s, the United States and the European Union gradually brought nominal interest rates down to zero, and monetary policy faced a serious problem worldwide: How can monetary policy ensure its effectiveness under zero interest rates?

Even if nominal interest rates fall to zero, there is still some work to be done. This is the position of the world's central banks (Bernanke 2020). If there is a floor to the nominal interest rate, it is the *real* interest rate that affects the economic behavior of firms and households. Because the real interest rate is the nominal interest rate minus the expected inflation rate, the real interest rate will fall if inflation expectations rise even if the nominal interest rate does not fall.

The idea that inflation/deflation is a monetary problem is ultimately based on the Quantity Theory of Money. The basic equation of the quantity theory of money is the "Cambridge Equation" by Marshall

$$M = kPy \tag{6.1}$$

or Fisher's "quantity equation"

$$MV = PQ \tag{6.2}$$

Here, M is the quantity of money (money stock), and P is the nominal price level. In the Cambridge equation, y is real GDP, to use today's terminology, and k *is a* positive coefficient called "Marshall's k". In the Fisherian quantity equation, Q is the real volume of transactions (flows), and V *is the* "velocity of money in circulation".

Following the textbook, assume that k and V are constant. Then, P *is* determined by the quantity of money, M. It varies proportionally to money stock. This is the basic proposition of the quantity theory of money. Prices are a monetary phenomenon. The final chapter of the international bestseller *Mankiw/Macroeconomics* (Mankiw, 2016) entitled "What we know and what we do not know" flatly states that "the most important lesson of macroeconomics" is that "the growth of the money stock is the ultimate determinant of inflation".

Insofar as the quantity theory of money holds true, deflation (a fall in price P) occurs because the increase in the money stock M does not keep pace with the increase in y, i.e., there is not an adequate supply of money. Theoretically, stopping deflation is easy. All one needs to do is to sufficiently increase the quantity of money, M. In fact, according to the quantity theory of money, increasing the quantity of money M is the only way to stop deflation.

The quantity theory of money has been around for a long time in economics. Ricardo (1817) strongly advocated the quantity theory at the beginning of the 19th century. In the postwar era, Milton Friedman's monetarism represented this old doctrine (Friedman, 1987). The modern microfounded macroeconomics, either neoclassical or neo Keynesian DSGE, has also come to believe in the quantity theory of money; in other words, in the "long run", nominal prices are determined by the quantity of money, M. In all the theoretical and simulation models, the Cambridge equation anchors price in the long run.

After the bursting of the bubble economy in the late 1990s, Japan's nominal interest rate, the target rate called the *call rate,* fell to zero. What is possible for monetary policy to do under the zero interest rate? Krugman (1998) examines monetary policy in a microfounded macromodel under the constraint of the zero interest rate known as the "liquidity trap". Krugman's paper entitled "It is Baaack: Japan's Slump and the Return of the Liquidity Trap" had a major impact on the academic community (Svensson, 2001; Eggertsson & Woodford, 2003; Bernanke et al., 2004; Auerbach and Obstfeld, 2005).

The Cambridge equation plays a central role in the Krugman model. The trick is that under the constraints of zero interest rates or a liquidity trap, the basic equation of the quantity theory of money, which is supposed to determine price P, actually becomes an *inequality* so that increasing M loses its power to raise P in a straight way. This is the fundamental problem of the liquidity trap. However, the Krugman model is a two-period model with not only the "current period" but also the "future". In the future, the quantity equation holds in its normal form. That is, the future economy is not caught in the liquidity trap. Therefore, future prices $P*$ are determined in proportion to the quantity of money $M*$ in the future.

The key word is then "expectations". If the central bank succeeds in creating expectations that $M*$ will increase enough, the real interest rates will fall at present because the expected rate of increase in price will rise. This will stimulate demand, even if the normal mechanism of lowering the nominal interest rate does not work. In this way, the economy can recover from recession. In the Krugman model, the purpose of increasing *M in the current period* is to create expectations for a significant increase in the money stock $M*$ *in the future.* In Krugman's own word, the central bank must "credibly promise to be incredible".

The model is based on consumer intertemporal utility maximization and, therefore, is taken by the profession as properly "microfounded". However, who holds the "expectation" that prices will rise in the first place? In the Krugman model, the representative consumer maximizes the intertemporal utility. Now, the only economic agents that exist in the model are the central bank and the representative consumer. The representative consumer understands everything the central bank does. They also share the same macromodel. In other words, both the central bank and the representative consumer have the same complete understanding of how the economy works. The key economic environment is the quantity equation of money or the Cambridge equation. The representative consumer in the model is a believer in the quantity theory of money! In this model, if money M^* will increase in the future, the rational consumer is expected to expect future price P^* to rise. This is exactly what central banks around the world were doing.

The question is whether the macromodel described above adequately models the real economy. Look again at Fig. 1.1b in Chap. 1. As shown in this figure, in all microfounded models, including the Krugman model, the micro agents acting rationally in the model share a single macromodel. This assumption is simply wrong. In reality, as shown in Fig. 1.1a, all the agents are optimizing in different "microcosms", i.e., under completely different micro constraints. Households and firms are more or less maximizing utility or profits under some constraints. However, each micro constraint is different and cannot be observed by a third party, including an economist analyzing the macroeconomy. There is no single microeconomic agent that optimizes taking one shared macroeconomy as a constraint. In particular, there is no household or firm that perceives to be constrained by stock of money or the Cambridge equation in making decisions. This is the fundamental difference between the so-called microfounded model and reality. A series of stochastic simulations by which Bernanke (2020) ascertains the efficacy of inflation targets also rests on the same incorrect assumption.

What the central bank can actually increase on its own initiative is the monetary base. On April 4, 2013, the Bank of Japan under the new governor Mr. Haruhiko Kuroda announced that it would double the amount of monetary base (hereafter referred to as "H") in two years. As shown in Fig. 6.4, H grew at a pace never before seen in the past.

Would such an increase in H affect consumers' expected inflation? The answer is no simply because consumers do not know what a monetary base means. In fact, the Bank of Japan regularly conducts a public opinion poll. As of March 2022, nine years after the Kuroda monetary easing started, 41.4% of people answered that they had never heard that the BOJ would ease money to achieve the inflation target of 2%. How can one's own actions be affected by changes in something one does not know about?

A former Federal Reserve chairman Alan Greenspan defined price stability as follows.

> This suggests, from a central banker's point of reference, an operating definition of price stability: Price stability obtains when economic agents no longer take account of the prospective change in the general price level in their economic decision-making.
>
> Since we cannot observe expectations directly, we look for proxies. If we believe that expectations are grounded in reality, then the relevant proxy is an index of the actual general price level (Greenspan, 1996, 1–2).

In the age of low inflation/deflation, people cease to pay attention to the prospective change in the general price level.

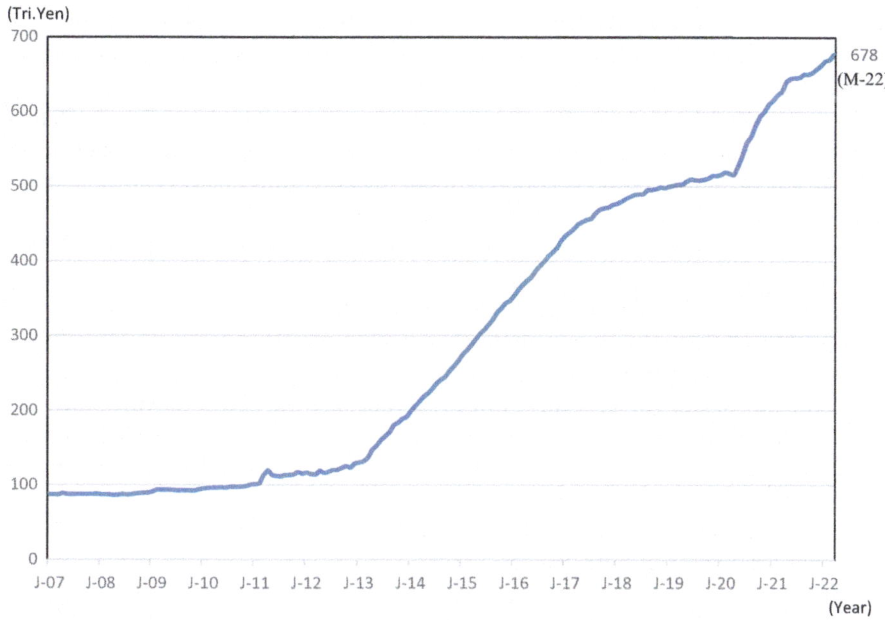

Fig. 6.4 Monetary base. *Note* Average amounts outstanding, Seasonally adjusted. *Source* Bank of Japan

The proposition that an increase in monetary base raises expected inflation is an impossible proposition from the start. Despite an unprecedented increase in money, the rate of inflation had never reached the BOJ's target of 2% for nine years. The experiment failed.

6.3 Determination of Price

Instead of depending on money stock, we must rethink how prices are determined. The general price, including the consumer price targeted by the central bank, is an index that is a weighted average of individual prices of goods and services. The weakness of the quantity theory of money, which appears to be impeccable, is that it relies on the equilibrium of demand for and supply of money, or the quantity equation, and simply states that the general price is proportional to the quantity of money *in equilibrium*. The process of how the actual price reaches that level is a complete "black box". Marshall was well aware of this problem. That is why he was hesitant, or rather almost opposed to applying the quantity theory of money to the deflation he faced during the Great Depression of the late nineteenth century (1873-96).[2] Modern microfounded macro models such as DSGE bypass the determination of price by way of representative consumer expectations.

To understand how price changes, we must go inside the "black box". That is, we must examine how the prices of individual goods and services are determined. Ball calls such an approach "accounting theory" and argues that it is an elementary mistake made by those who do not know economics and confuse relative and nominal prices (Ball, 2006). However, as we explain below, it is actually economists such as Ball who make an elementary mistake.

Suppose that there is a change in the economy that causes the "relative price" of good A to rise potentially; for example, demand for good A increased, or the cost of production of good A rose. In theory, it is certainly possible that the nominal price of good A *declines*, and yet its relative price rises because the extent to which the price of A declines is less compared to the declines of prices of other goods and services. However, this is not what we normally observe in reality. In the real economy, contrary to Ball's argument, a "common sense" change occurs *independently of the money stock*; for example, if the demand for good *A* increases, the (nominal) price of A rises, and if the demand falls, the price falls.

Unlike asset prices and commodities, the actual increase in the price of most manufactured goods and services depends on the (explicit or implicit) "negotiations" between the supplier of good A and the "economic agents involved". The concept of *fairness*, as stressed by Hicks (1974) and Okun (1981), plays a major role in the pricing of good A by the supplier who seriously considers its "stakeholders". A baker who raises the price of bread with ease will lose many customers forever if it is not seen as fair.

Fairness plays an important role in reality. Kahneman, Knetsch and Thaler (1986), in their paper entitled "fairness as a constraint on profit seeking", demonstrate it through telephone surveys.

> The study was concerned with scenarios in which a *firm* (merchant, landlord, or employer) makes a pricing or wage-setting decision that affects the outcomes of one or more *transactors* (customers, tenants, or employees). The scenario was read to the participants, who evaluated the fairness of the action as in the following example:
>
> Question 1. A hardware store has been selling snow shovels for $15. The morning after a large snowstorm, the store raises the price to $20. Please rate this action as:
>
> Complete Fair Acceptable Unfair Very Unfair

The two favorable and the two unfavorable categories are grouped in this report to indicate the proportions of respondents who judged the action acceptable or unfair. In this example, 82 percent of respondents ($N = 107$) considered it unfair for the hardware store to take advantage of the short-run increase in demand associated with a blizzard (Kahneman et al., 1986, 729).

[2] For example, his letter to Bonar dated 6 March 1899 (Pigou, A. ed. 1925, *Memorials of Alfred Marshall*, London: MacMillan, 375) reads as follows. Emphasizing the importance of the real economy, Marshall wrote "I am never weary of preaching in the wildness 'the only very important thing to be said about currency is that it is not nearly as important as it looks'.".

Any market in which individual price is determined and fairness is involved is a *local market* (the "microcosm" described earlier). For economic agents involved in determining prices/wages locally, the money stock, let alone the monetary base, is not an economic variable that they take into account in their decision-making. As already mentioned, the quantity theory of money does not have a place in the micropricing process from beginning to end.

Now, the "fairness" in determining the price of an individual good or service is "fairness" relative to the prices of other goods and services that are a given. Kahneman et al. (1986) states the following:

> A central concept in analyzing the fairness of actions in which a firm sets the terms of future exchanges is *reference transaction*, a relevant precedent that is characterized by a reference price or wage and by a positive reference profit to the firm (Kahneman et al., 1986, 729).

Without the assumption that the prices of other goods/services are given, the very playing field for determining individual goods/services is lost. Namely, it means the loss of the very foundation of the monetary economy itself. In the monetary economy, what changes the "relative prices" of individual goods and services is actually a change in the "absolute prices". The general price level is nothing but the result of a weighted average of such individual nominal prices of goods and services that are determined locally.

Once again, without the quantity theory of money, how is the price of goods and services determined? Kalecki provided a theoretical framework for this question (Kalecki, 1954). Kalecki suggested that when analyzing the macroeconomy, it is necessary to make a clear distinction between "prices determined by production costs" on one hand and "prices determined by demand" on the other; he proposed a two-sector approach.

> Short-term price changes may be classified into two broad groups: those determined mainly by changes in cost of production and those determined mainly by changes in demand. Generally speaking, changes in the prices of finished goods are 'cost-determined' while changes in the prices of raw materials inclusive of primary foodstuffs are 'demand-determined.' The prices of finished goods are affected, of course, by any 'demand-determined' changes in the prices of raw materials but it is through the channel of *costs* that this influence is transmitted.
>
> It is clear that these two types of price formation arise out of different conditions of supply. The production of finished goods is elastic as a result of existing reserves of productive capacity. When demand increase it is met mainly by an increase in the volume of production while prices tend to remain stable. The price changes which do occur result mainly from changes in costs of production.
>
> The situation with respect to raw materials is different. The increase in the supply of agricultural products requires a relatively considerable time. This is true, although not to the same extent, with respect to mining. With supply inelastic in short periods, an increase in demand causes a diminution of stocks and a consequent increase in price. This initial price movement may be enhanced by the addition of a speculative element (Kalecki, 1954, 11).

The prices of manufactured goods and services that Kalecki argues are determined based on "production costs" also change in part, reflecting changes in demand, namely, the markup rate changes. Nevertheless, Kalecki's two-sector approach was basically correct and adopted by Keynesians. Specifically, following the lead of the classic empirical study by Hall and Hitch (1939), markup pricing based on production costs has been repeatedly confirmed (Godley and Nordhaus 1972; Coutts, Godley and Nordhaus 1978). In passing, Godley and Nordhaus emphasize that mark-up pricing is based on *normal* production cost rather than production cost every moment. This result is consistent with the kinked individual demand curve explained in Chap. 3.

As Kalecki argues, the price of primary commodities such as agricultural products, oil and other minerals is basically determined by the market so that supply and demand are matched as in the textbooks. For many commodities, the futures market exists where speculation and therefore expectations play a major role.

However, the pricing of manufactured goods and services is quite different from the pricing of these primary products. Prices are set by producers (not the market), and in doing so, the producer's primary concern is the "cost of production."[3]

Production cost-based pricing is confirmed by the study of fairness through telephone surveys:

> Question 7. Suppose that, due to a transportation mixup, there is a local shortage of lettuce and the wholesale price has increased. A local grocer has bought the usual quantity of lettuce at a price that is 30 cents per head higher than normal. The grocer raises the price of lettuce to customers by 30 cents per head.
>
> (N = 101) Acceptable 79% Unfair 21%
>
> Question 8. A landlord owns and rents out a single small house to a fixed income. A higher rent would mean the tenant would have to move. Other small rental houses are available. The landlord's costs have increased substantially over the past year and the landlord raises the rent to cover the cost increases when the tenant's lease is due for renewal.
>
> (N = 151) Acceptable 75% Unfair 25%
>
> (Kahneman et al., 1986, 732–733).

Their conclusion summarized as a proposition reads as follows:

PROPOSITION 3: Price changes will be more responsive to variations of costs than to variations of demand and more responsive to cost increases than to cost decreases (Kahneman et al., 1986, 738).

Unfortunately, this seems all but forgotten now. It is surprising that when deflation/inflation is so much of a problem, economists are trying to analyze changes in prices without having the "right theory" about how prices are determined.

Fluctuations in demand are essentially absorbed by changes in the "quantity of production" rather than prices, although they may affect the mark-up rate.[4] This is nothing but Keynes's principle of effective demand. Kaldor said the following.

---

[3] The prices are essentially determined based on the cost of production is readily apparent with a little careful observation of actual price changes. For example, in November 2018, the Salt Business Center in Japan announced that the price of salt would be raised by 6–25% starting in April 2019. The reasons cited by the Center for this change were the rising price of coal, the fuel needed for production, and the rising price of sun-dried salt imports, as well as the price of packaging materials and logistics-related charges, which the Center said producers of salt could no longer absorb "on their own". This is a typical reason cited by producers/firms when they raise prices. They do not know the definition of the monetary base or money stock, so of course they would not mention it.

[4] Recent analyses of micro prices (such as POS data) emphasize "special sales," which are important for super market management strategies, but are a minor issue for macroeconomics.

> The important conclusion is that the signal that causes an economic "agent" to do something different—produce more or produce less or switch his manufacturing facilities from some varieties to others—is always a quantity signal, not a *price* signal. Prices are set by the producers on normal costs of production (or rather, on the costs calculated by reference to normal utilization of capacity) including a customary percentage added for profit; and within limits, the producer will not change his price as a result of a faster (or slower) increase in orders, unless the increase in demand signaled to him is so large that he cannot cope with it without disappointing his regular customers, or else the fall in demand is so large that it causes him to incur standstill costs due to lack of orders (keeping workers and machinery idle), in which case he might try to avoid some of these untoward consequences by some temporary price concession or a price concession that is not formally announced but that he is willing to concede in bargaining. In any case, in the actual adjustment of supply and demand, prices play only a very subordinate role, if any (Kaldor, 1985, 24–25).

For goods and services that are manufactured, the principle of effective demand applies where "quantity" determines "quantity". What determines the *prices* of manufactured goods and services? As already mentioned, according to Kalecki, the price P of manufactured goods and services is determined by the producer by adding the profit margin to the production cost. The most important factor in determining the prices of manufactured goods and services is the production cost, even though the markup may vary slightly. This can be expressed in the following equation. That is, the price P is equal to

$$P = \text{"markup"} \times \text{"production cost per unit of product (service)"} \tag{6.3}$$

Markups are strategically determined by producers, taking into account the price elasticity of demand, and the level of markup varies by industry, region, and company. However, as Kaldor says, once a markup level is set, it does not fluctuate much in the short term. Therefore, the main cause of price fluctuations is basically the production cost per unit of product.

Production costs consist of labor costs and raw material costs (including energy). If the wage is W, the labor input is L, the price of raw materials is P_R, and the amount of raw material input is R, and if Y (goods and services) is produced with L and R, then the "cost of production per unit of product", $COST$ is

$$COST = \frac{WL + P_R R}{Y} \tag{6.4}$$

In the case of Japan, almost all raw materials (including oil as a source of energy) are imported. Therefore, the yen-denominated price P_R of raw materials can be expressed as the international price P_R * in dollars and exchange rates e (1 dollar = e yen).

$$P_R = eP_R^* \tag{6.5}$$

Incidentally, in the production of final goods, some manufactured goods such as H-shaped steel and polypropylene produced by other firms are used. This creates the dynamics of individual prices from "upstream" to "downstream", and accordingly the dynamics of the general price, namely, inflation and deflation. We will analyze this important problem in the next section.

The production cost per unit of product *COST* can be rewritten as

$$COST = \frac{WL + eP_R^* R}{Y} = \left(\frac{W}{l}\right) + \left(\frac{eR_R^*}{r}\right) \tag{6.6}$$

Here, l on the right side is

$$l = \frac{Y}{L} \tag{6.7}$$

and represents labor productivity. In the same way,

$$r = \frac{Y}{R} \tag{6.8}$$

defines raw material productivity.

In summary, the cost of production, which drives prices, depends on five factors. First, an increase in nominal wages W *is* a factor that raises prices through an increase in labor costs.[5] However, the pressure for higher wages is mitigated by an increase in labor productivity l. An increase in wages will not create upward pressure on prices if productivity l rises by the same amount. Put another way, if nominal wages do not rise as much as labor productivity rises, this is a factor that depresses prices. We will shortly show that this factor was significant during Japan's deflation since the late 1990s.

Along with labor costs, raw material cost changes prices. Japan imports some agricultural products, such as wheat, and almost all of its raw materials. Hence, the yen-denominated price of raw materials P_R is affected by both the international primary commodity prices P_R* (in dollars) and the exchange rate e.

In the long run, purchasing power parity (PPP) theory holds that the nominal exchange rate is determined by the level of domestic and foreign prices. In fact, the long-term trend of the yen's appreciation from ¥360 to the dollar, which lasted for more than twenty years after World War II, to the ¥120 level after the Plaza Accord in the fall of 1985 was basically the result of the rise in labor and raw material productivity (the rise of l and r) in Japan, which caused wholesale prices (currently corporate prices) to fall at a higher pace than in other countries (Yoshikawa 1990).

[5] Peneva and Rudd (2017) report, however, that the passthrough of labor costs to price inflation has weakened in the U.S. in recent years.

However, contrary to the long-run causal relationship in which prices determine the exchange rate, the exchange rate has a large impact on the domestic price level in the short run through changes in the prices of imported raw materials. Specifically, when the yen depreciates (e increases), domestic prices of raw materials rise, while if the yen appreciates (e decreases), domestic prices fall. For example, in both 2008 and 2011, the price of crude oil rose to an annual average of over \$95 a barrel. However, while the exchange rate was 103 yen to the dollar in 2008, the yen appreciated 79 yen to the dollar in 2011. As a result, the yen-denominated price of crude oil remained at 7,585 yen per barrel in 2011 (up to only 8.8% from the previous year) compared to 10,293 yen per barrel in 2008 (up to 21% from the previous year). The CPI, including energy, increased by 1.4% in 2008, whereas it decreased by 0.3% in 2011. Although how much of the cost increase is passed on to product prices depends on many factors, it is clear from the experience of 2008 and 2011 that this mechanism is alive and well in the open economy.

From the equation that describes how the "level" of price P is determined, we can easily derive an equation that describes how the rate of change π of price P is determined. The rate of change in price π

$$\pi = \frac{\Delta P}{P} \tag{6.9}$$

which is nothing but inflation (where π is positive) or deflation (where π is negative) is determined as follows:

$$\pi = \alpha(w - g_l) + (1 - \alpha)(\hat{e} + \pi_R^* - g_r) \tag{6.10}$$

Here, α is the share of labor cost in production cost (thus, $1 - \alpha$ is the share of raw material cost in production cost). w is the rate of change in nominal wages W (up if positive, down if negative), and π_R^* is the rate of change of the international price of primary products, ${P_R}^*$. The rates of change in labor productivity l and raw material productivity r are g_l and g_r, respectively. $\hat{e}$ is the rate of change in exchange rate e (a positive value of e indicates a depreciation of the yen, while a negative value of e indicates an appreciation of the yen).

The BOJ's September 21, 2016 "summary review" of monetary policy over three and a half years cited "falling oil prices" as the reason why prices did not rise as initially (namely, April 2013) expected. As Eq. (6.10) shows, this in itself is correct. However, this is rejected by the economics–monetarism–on which the central bank itself relies.

Milton Friedman, the proponent of monetarism or the leader of the modern quantity theory of money, rejected the argument that higher oil prices caused or even exacerbated inflation, saying

> Not at all. It is essential to distinguish changes in relative prices from changes in absolute prices. The special conditions that drove up the prices of oil and food required purchasers

> to spend more on them, leaving less to spend on other items. Did that not force other prices to go down or to rise less rapidly than otherwise? Why should the average level of all prices be affected significantly by changes in the prices of some things relative to others? Thanks to delays in adjustment, the rapid rises in oil and food prices may have temporarily raised the rate of inflation somewhat. In the main, however, they have been convenient excuses for besieged government officials and harried journalists rather than reasons for the price explosion.
>
> The basic source of inflation is the faster growth in the quantity of money than in output. (Friedman, 1974, 73).

What a loose logic! Consumers may not necessarily spend less on items other than oil and food to keep the level of average price intact. Anyway, we never hear of declines of prices of many consumables when oil price skyrocketed. The problem Japan faced around the year 2000 was deflation, not inflation, but the same logic should apply in so far as one believes in monetarism, or any doctrine that price is determined by money stock.

In fact, contrary to Friedman's assertion, Brown and Ozga (1955), analyzing the eighty-year record of British deflation/inflation, found that the most important determinant was the terms of trade, which was in turn basically driven by the prices of raw materials. Today, the U.S. and EU face the danger of high inflation; in March 2022, the rate of increase of consumer price reached 8.5% in the U.S. and 4.9% in Germany. Meanwhile, the price of oil (WTI) has increased from 3.9 US$ on April 20th 2020 to 106.05 US$ on February 28th 2022. There is no denying that this price hike has contributed to recent inflation. It wiped out confidence in the microfounded DSGE, which was supposed to guide the central bank to the inflation target of 2%.

Now, the price of raw materials has a significant impact on domestic prices, but since the prices of many raw materials are competitively determined in the international market, the ups and downs should have an almost universal effect on all countries, apart from exchange rate movements. Therefore, the answer to the question of why Japan alone had lapsed into deflation for nearly two decades must be sought elsewhere.

We focus on nominal wages as a key variable in Japan's deflation (Ushijima, 2004; Yoshikawa, 2013). As already explained, the trend of nominal wages is an important variable that determines the movement of nominal prices. As seen from Fig. 6.5, the decline in nominal wages in Japan is highly unusual on an international scale. Nominal wages, which play a major role in determining production costs, have fallen in Japan alone.

The reason why advanced countries did not experience deflation after WWII is that, in contrast to the prewar period, nominal wages did not fall after the war; see Akerlof, Dickens and Perry (1996) for the U.S. In other words, the downward rigidity of nominal wages was the "deflation stopper". In Japan, however, this deflation stopper was lifted from the end of the 1990s. We will consider wages in Sect. 6.5. Before doing so, we next turn to the relationship between individual prices and the aggregate price index.

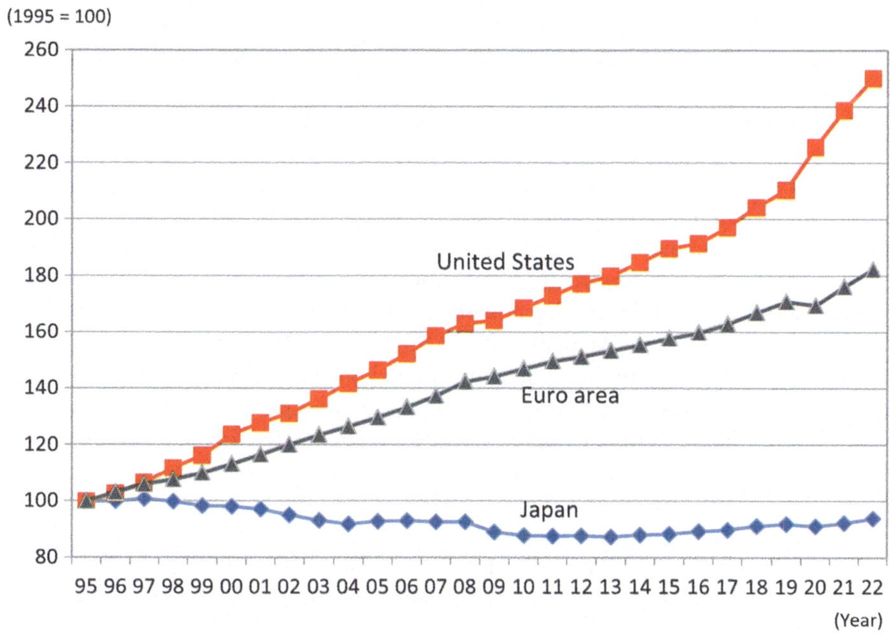

Fig. 6.5 International Comparison of Nominal Wages. *Note* Indexed with 100 representing the level in 1995 by using the percent changes from the previous year in Annex Table 17, "Compensation per employee". *Source* OECD, *Economic Outlook*, Volume 2021, Issue 2

6.4 Individual Prices and Aggregate Price Index

In macroeconomic theory, prices are said to be "sticky." In fact, in modern dynamic stochastic general equilibrium (DSGE) models, monetary policy is effective to the extent that prices are sticky. A number of theories attempting to explain sticky prices have been proposed: the Taylor–Calvo model of desynchronized staggered wage/price changes and menu cost models, to name a few. However, as is customary in microfounded macroeconomics, these models assume that all firms are "symmetric" and, by considering the optimal behavior of representative firms, provide a framework for understanding inflation/deflation. Such a theory can shed some light only on cross-sectional patterns of individual prices but does not provide an adequate answer to the question of how aggregate prices fluctuate. Inflation/deflation is a change in general prices, after all. Standard theoretical models of individual price changes cannot provide an effective analysis of inflation or deflation.

Based on microfoundations, the standard framework for understanding the role of monetary policy is the New Keynesian Phillips curve (NKPC). As mentioned in Sect. 6.2, the key property of the NKPC is that inflation is primarily a forward-looking process. That is, expectations on future inflation largely determine current inflation. This justifies recent emphasis on expectations management and communications as tools of monetary policy. A large amount of literature exists on the NKPC. However, after a lengthy survey of the literature, Mavroeidis, Plagborg-Møller, and Stock (2014) reached quite a disappointing conclusion. Namely, their major finding was that estimation of the NKPC using macroeconomic data is subject to a severely weak instrument problem. Indeed, they found that "the evidence is consistent both with the view that expectations matter a lot, as well as with the opposite view that they matter very little". They thus concluded that identification of the NKPC is too weak to warrant research on conceptually minor extensions. The traditional analysis based on macroeconomic data has clear limitations.

Meanwhile, recent empirical studies on micro price-setting, as surveyed by Klenow and Malin (2011), have uncovered the hitherto unknown dynamics of micro prices (Sudo, Ueda, and Watanabe (2014) for Japan). Bils and Klenow (2004), for example, by examining the frequency of price changes for 350 categories of goods and services, demonstrated that half of the prices remain valid for 5.5 months or less. Their findings seemed to suggest that individual prices are not actually rigid. There are substantial differences across goods; however, the prices of raw materials and foodstuffs are flexible, whereas those of services are less flexible. This fact is well known and is in good accordance with Kalecki (1954)'s two-sector framework explained above. Today, central banks are committed to inflation targeting with respect to the "core" CPI, which excludes the prices of foodstuff and energy. Thus, they focus on what Hicks called the "fixed price sector".

In what follows, we explore the dynamics of individual prices. In particular, we focus on interactions of prices with leads and lags. They play a significant role in explaining the behavior of the aggregate price index. Thus, to understand inflation and deflation, it is essential to explore the mechanism of interactions among prices. Resorting to a new method, we show that irrespective of the sources of shocks, there exists a robust flow of changes in domestic prices from upstream to downstream. Moreover, we demonstrate that *prices change in clusters*. Overall, our analysis shows that the inertia arising from input/output linkages in production basically explains the behavior of aggregate prices.

As a first step in understanding actual price movements, we examine the relationship between the rate of change in the consumer price index (CPI) and the rates of change in individual prices that make up CPI. The relationship between the rate of change in the CPI and the rates of change in individual prices is by definition as follows.

$$\pi(t) = \sum_{\alpha=1}^{n} g_\alpha \pi_\alpha(t) \tag{6.11}$$

Here, g_α is the share of prices of goods/services $\alpha(\alpha = 1, \ldots, n)$ in the consumer price index. Then, the autocorrelation of the rate of change in consumer prices can be expressed as follows.

$$\langle \pi(t_0)\pi(t_0+t)\rangle_{t_0} = \sum_{\alpha} g_\alpha^2 \langle \pi_\alpha(t_0)\pi_\alpha(t_0+t)\rangle_{t_0} + \sum_{\alpha \neq \beta} g_\alpha g_\beta \langle \pi_\alpha(t_0)\pi_\beta(t_0+t)\rangle_{t_0} \tag{6.12}$$

In other words, the autocorrelation of the rate of change in CPI is the sum of (1) the autocorrelations of individual prices (the first term of the right-hand side of Eq. (6.12)) and (2) the interactions between different individual prices (the second term).

As is typical of Mankiw's (1985) menu-cost model, many theoretical models attempt to capture general price movements in terms of "representative" micro price movements. The reason why aggregate price is "rigid" is that the movements of individual prices determined by firms are "rigid". However, the autocorrelation of the aggregate price turns out to be completely different from the autocorrelations of individual prices, as will be shown below.

The autocorrelation coefficients of the rates of change in the prices of the 47 goods and services that make up the consumer price index are shown in Fig. 6.6a. Although the autocorrelation coefficients differ for each good, they all show a common pattern of monotonically decreasing up to 12 months and then moving horizontally at a small value close to zero thereafter. Figure 6.6a is a weighted average of the autocorrelations of the 47 individual price changes.

The pattern in Fig. 6.6a can be naturally explained as follows. Suppose that the rate of change of individual price $p_\alpha(t)$ *relative to the previous month* is.

$$\log p_\alpha(t) - \log p_\alpha(t-1) = \varepsilon_t. \tag{6.13}$$

Namely, if it is a random walk as in (6.13), then the rate of change in "year-on-year" is expressed by the following Eq. (6.14).

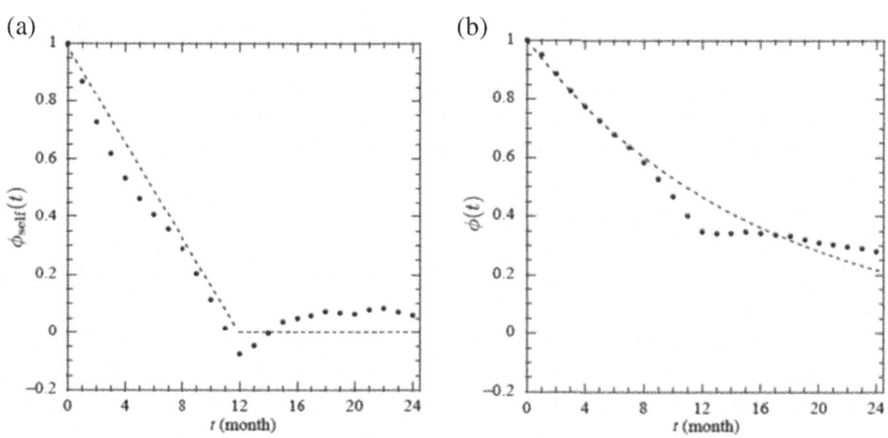

Fig. 6.6 The Autocorrelation of Prices. (a) The autocorrelation of individual prices. (b) The autocorrelation of consumer prices. *Source* Kichikawa et al. (2018)

$$\pi_\alpha(t) = \log p_\alpha(t) - \log p_\alpha(t-12) = \epsilon_t + \epsilon_{t-1} + \cdots + \epsilon_{t-11} \tag{6.14}$$

The autocorrelation is then

$$\phi_\alpha(t) = 1 - \frac{t}{12} \quad (0 \le t \le 12) \tag{6.15}$$

In other words, as shown in Fig. 6.6a, it decreases monotonically until 12 months and then falls to almost zero. In passing, the random walk of individual price movements, as shown in Eq. (6.13), indicates that the micro shocks are permanent.

The autocorrelation of the year-on-year rate of change in the consumer price index shown in Fig. 6.6b is quite different from that of individual prices. It can be approximated by an exponential function.

$$\phi(t) = \exp(-t/\tau) \tag{6.16}$$

While individual price changes have a relatively short-term memory of 12 months or less, the rate of change in the consumer price index, namely, the weighted average of individual price changes, has a much longer memory. In other words, it is *inflexible*.

The difference between the two autocorrelations[6] arises from the second term on the right-hand side of Eq. (6.12), namely, the *interactions of individual prices*. Therefore, to understand the "rigidity" in changes in aggregate price index such as CPI, it is necessary to know the interactions of a number of individual prices rather than the pricing of individual prices.

Kichikawa et al. (2020) examines these interactions of individual prices using a technique called *complex Hilbert principal component analysis*. This is an extension of principal component analysis that uncovers "hidden common factors" that produce correlations among numerous time series data. Ordinary principal component analysis is a standard method used in many fields, and we used it to analyze business cycles in Chap. 4. Complex principal component analysis may sound peculiar at first, but phases of complex variables nicely capture leads and lags.

The data used are 80 monthly changes in Japan's individual prices, consisting of import prices (ID = 1–10), corporate prices (ID = 11–33), and consumer prices (ID = 34–80), for the period from January 1985 to December 2016. In addition, seven macro variables, namely, the yen/dollar exchange rate (ID = 81), the leading/coincident/lagging indicators of business cycles (Cabinet Office, ID = 82, 83, 84), money stock (ID = 85), monetary base (ID = 86), and nominal wages (manufacturing, Ministry of Health, Labor and Welfare, ID = 87), are used.

After "complexifying" these 87 monthly data, all of which are normalized annual rates of change from the same month in the previous year, by the "complex Hilbert transform", the correlation matrix of the 87 variables is calculated to obtain the eigenvalues. By *random matrix theory*, two eigenvalues are found to be significant.

[6] A test of statistical significance of the difference between the two autocorrelation functions shown in Fig. 6.6a, b can be found in Kichikawa et al. (2018).

Random matrix theory provides a systemic method of significance test for eigenvalues in principal component analysis; see Aoyama et al. (2017) for complex Hilbert principal component analysis and random matrix theory.

The two significant complex eigenvalues, which are 87 dimensional vectors, represent the "hidden main causes" that generate the correlations between the 87 time series. Movements of the aggregate price index, such as CPI, arise from the correlations of individual prices.

In Fig. 6.7a, b are the 87 elements that make up the two significant eigenvectors. The numbers in the figures are the IDs of the components explained above. Because a component of the eigenvector is a complex number, it has the absolute value and the phase. The absolute value on the vertical axis in Fig. 6.7 expresses how much impact each variable corresponding to the ID number has on the collective movements of prices, while the phase on the horizontal axis expresses the leading/lagging relationship of each variable (0 on the left side of the figure is leading, and 2π on the right side is lagging). The horizontal dashed line in the figure indicates that if the absolute value of the variable is below this line, the contribution of this variable to the collective movement is not significantly different from zero.

In the first eigenvector, the absolute value of the exchange rate (ID = 81) is large and leads other variables. Additionally, the absolute values of import prices and corporate prices are large and lead other variables. The absolute value decreases from corporate prices (finished product) to consumer prices. The first eigenvector thus captures the impulse of the exchange rate and the diminishing effects from import prices to corporate prices and then eventually to consumer prices with delay and decay.

In contrast, the second eigenvector captures the movement of domestic prices (corporate/consumer prices) driven by the domestic economic trend (ID = 82, 83, 84) as an impulse. These two eigenvectors capture the "systematic collective movements" of individual prices and thus their weighted average and aggregate price index.

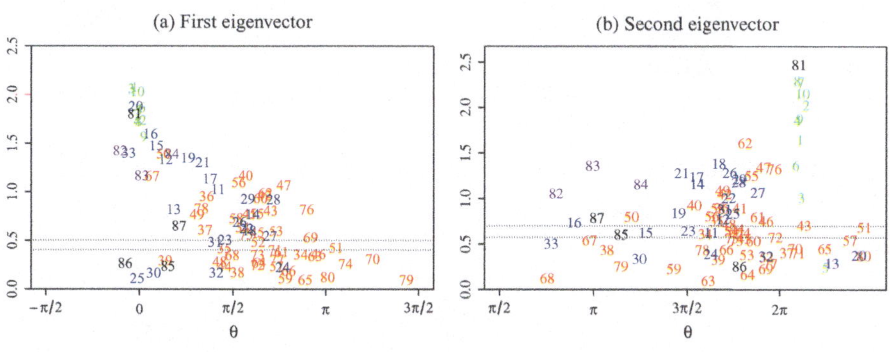

Fig. 6.7 The Components of Eigenvectors Associated with the Largest and the Second Largest Eigenvalues. (a) The first eigenvector. (b) The second eigenvector. *Note* The number shown in the figures is the ID of each variable. The absolute value of each complex variable is vertically shown, whereas horizontally, the phase is shown. *Source* Kichikawa et al. (2020)

The macro variables that have significant impacts on price movements are the exchange rate and the business cycle indicators. Nominal wages (ID = 87) also have a significant impact. In contrast, neither money stock nor monetary base, which is supposed to play a central role in standard theoretical models, has a significant impact; the absolute values of ID=85, 86 are below the horizontal dashed line.

Although the two eigenvectors capture different kinds of impulse-generating price movements, the phase of individual prices, i.e., the leading/lagging relationship, is basically the same for domestic price movements alone. This suggests that there is a single mechanism for the interactions of domestic prices regardless of the nature of the macro shocks —— exchange rate changes or domestic economic conditions —— that generate price fluctuations.

In fact, domestic price spillovers are not random. Individual prices form *clusters*, and price movements spill over through these clusters. Kichikawa et al. (2020) identifies these clusters by using a percolation model. The results of the analysis are shown in Fig. 6.8. Here, we focus on the period after the Plaza Accord in the fall of 1985 and the global recession triggered by the collapse of Lehman Brothers in September 2008. In both periods, prices abruptly changed in a short period of time.

After the Plaza Accord reached the Plaza Hotel in New York in the fall of 1985, the yen sharply appreciated from ¥240 per dollar at the beginning of the year to 120 yen per dollar in less than two years. As a result of the yen's appreciation, import prices and corporate prices for materials fell rapidly (Fig. 6.8b). However, the decline in prices has not spread to consumer prices. This is basically the movement explained by the first eigenvector.

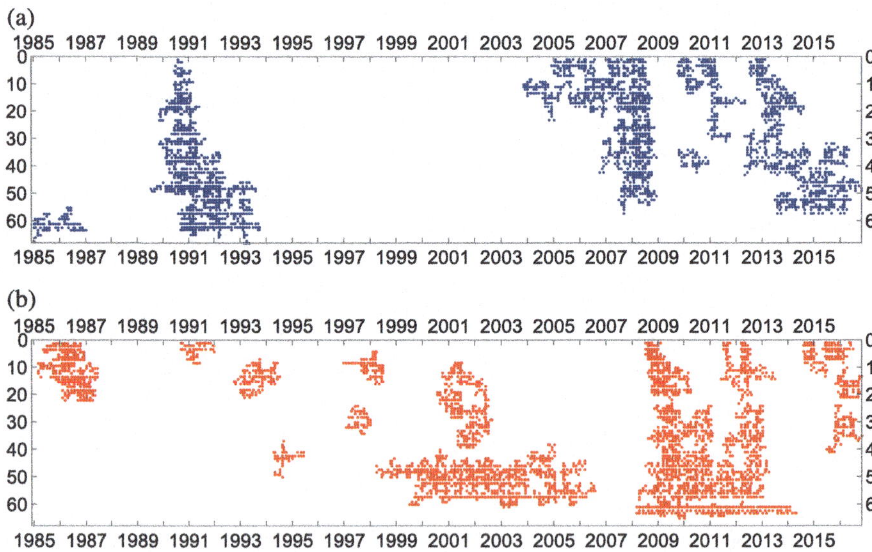

Fig. 6.8 Price Clusters. (a) Prices going up by sector. (b) Prices going down by sector. *Note* Fig. 6.8a and b show the indices of individual prices that significantly rise, (a) in blue, and decline, (b) in red, respectively. In both figures, the IDs of individual prices (1–80) are shown on the vertical axes from upstream (top = 1) to downstream (bottom = 80). *Source* Kichikawa et al. (2020)

After a temporary recession caused by the rapid appreciation of the yen, the Japanese economy experienced a "bubble economy" driven by domestic demand. Prices stopped falling, and prices of imported materials began to rise in 1987. The price rise spread to consumer prices in 1990–93. From 1991, when the bubble burst and the recession began, material prices fell first, but consumer prices *rose* even as the bubble burst.

The Japanese economy had been in deflation since the end of the 1990s, but in 2007, with the economic expansion and the rise in international prices of crude oil, wheat and other commodities, most prices rose with spillovers from material to consumer prices. The end of deflation was clearly in sight in the middle of 2008 (Fig. 6.8a). However, with the collapse of Lehman Brothers in September, the world economy abruptly fell to a "once-in-a-century" recession. The tide turned quickly, and prices of almost everything from materials to final consumer goods suddenly stopped rising and started falling (Fig. 6.8b).

The changes in individual prices in Fig. 6.8 show that individual prices change from "upstream" (raw materials) to "downstream" (final consumer goods) through the clusters. This is because, as pointed out by Kalecki, the prices of most manufactured goods and services are determined based on production costs and thus reflect the input/output chains. The movement of clusters clearly seen in Fig. 6.8 suggests that expectations do not play a major role in price movements, as there is no room for expectations in the transition from upstream to downstream clusters shown in Fig. 6.8.

6.5 Wages

The prices of many goods and services are basically determined on the basis of production costs. Among these costs, nominal wages have a particularly large impact on the cost of production.

Hicks' *The Crisis of Keynesian Economics* (Hicks, 1974, 59), which discusses Keynesian economics, makes the following interesting point. Anyone who reads Keynes's *General Theory* will find it strange that Keynes defines all quantities of income, consumption, investment, and even monetary quantities as "wage units"; Nominal values are all deflated by nominal wages. According to Hicks, this is because Keynes had an idea that could be called "the Wage Theorem". The Wage Theorem asserts that if nominal wages rise, all nominal prices will rise proportionally. As this "theorem" holds, even nominal quantity of money changes in proportion to the nominal wage. Thus, the quantity of money is a passively determined endogenous variable. Of course, this does not hold as a "theorem" in all countries at all times. However, Keynes believed that this was largely true in Britain in the 1930s.

If not as strong as the "wage theorem", it is true that in the postwar economies of the advanced nations, wages have had a significant impact on prices. The crucial question is then how wages are determined.

Neoclassical economics holds that real wages are essentially determined in the same way as the price of ordinary goods and services. Namely, wages are determined at an equilibrium level that matches the supply of and demand for labor. If for whatever reason wages are higher (lower) than this equilibrium level, an excess supply of labor (demand) is created. In fact, Pigou, the master of neoclassical economics and Keynes's greatest opponent in his *General Theory*, analyzed the British economy in the 1920s and 1930s from this orthodox position.

In the 30 years leading up to the outbreak of the First World War in 1914, unemployment in Britain only exceeded 10% in one year, 1886. The second highest was 8% in 1908. In contrast, after the end of the War, the unemployment rate did not fall below 9.5% for a single year between 1921 and 1939. How to explain the abnormally high unemployment rate was naturally a major challenge for the British economists. Pigou (1927) suggested that the main reason for the abnormally high unemployment rate was that real wages were too high relative to the equilibrium level after the War. The same idea that real wages were too high relative to the equilibrium level became popular when the unemployment rate rose globally after the oil embargos in the 1970s. Such unemployment was called "classical unemployment" (Malinvaud, 1982, Bruno sad Sachs, 1985).

In the 1920s, Pigou's analysis was immediately criticized. Clay (1928) examined the relationship between nominal wages and unemployment by sector rather than for the British economy as a whole and found that unemployment was lower in sectors with higher nominal wages, and conversely, unemployment was higher in sectors with lower nominal wages. This fact clearly shows that we cannot talk about wages as the price of uniquely defined "labor".

What should we think about wages? Wages are certainly the price of labor. However, there are actually two different types of "labor". We must start by confirming this fact. The fact that there are literally many different kinds of "types" or "qualities" of labor is discussed in Chap. 3 in relation to matchings of workers and firms. Here, the point is that even when we take a bird's eye view of the labor market, we must consider at least two types of labor. Hicks wrote about his first book, *The Theory of Wages* published in 1932, that although he had changed his mind about many things he wrote there, he still thought the following point was true.

> It is the distinction between casual employment, the single job implying no durable relationship, and regular employment, in which people work together and go on working together (Hicks, 1974, 64).

Hicks made the following remark on formal employment based on a long-term relationship.

> Now it is necessary, purely on grounds of efficiency, in regular employment, that both parties, employer and employed, should be able to look forward to some durability in their relationship. Yet if the worker is to be free to move (and if he is not free to move it is semi-slavery) there can be no such reliability unless there is contentment, or at least some degree of contentment. So it is necessary for efficiency that the wage-contract should be felt, by both parties but especially by the worker, to be *fair* (Hicks, 1974, 64).

It is fairness that plays an important role in determining the wages of "regular" labor. In other words, the labor market for formal workers is not quite a "market" in the usual sense of the word. Solow (1990) called the labor market a "*social institution*". It applies only to informal labor that wages are determined in equilibrium where supply and demand are equal, similar to the price of apple. However, people and apples are essentially different after all. Therefore, it may be actually an exaggeration to assume that even wages of irregular/secondary workers are determined in the market because the analysis of fairness based on surveys suggests entitlement on the part of casual workers:

> The role of prior history in wage transactions is illustrated by the following pair of questions:
>
> Question 2A. A small photocopying shop has one employee who has worked in the shop for six months and earns $9 per hour. Business continues to be satisfactory, but a factory in the area has closed, and unemployment has increased. Other small shops have now hired reliable workers at $7 an hour to perform jobs similar to those done by the photocopy shop employee. The owner of the photocopying shop reduces the employee's wage to $7.
>
> (N = 98) Acceptable 17% Unfair 83%
>
> Question 2B. A small photocopying shop has one employee... [as in Question 2A]...The current employee leaves, and the owner decides to pay a replacement $7 an hour.
>
> (N = 125) Acceptable 73% Unfair 27%
>
> The current wage of employee serves as reference for evaluating the fairness of future adjustments of that employee's wage - but not necessarily for evaluating the fairness of the wage paid to a replacement.
>
> (Kahneman et al. 1986, 730).

All theoretical macro models, whether endogenous growth models or RBCs, are general equilibrium models in which (real) wages are determined in such a way that labor L *is* fully employed. However, this framework does not capture reality at all. As Hicks, also quoted in Chap. 1, says, "The Walras model, which has become so popular with modern and especially mathematical economists, requires a form of organization which has rarely if ever existed".

The idea that wages are determined not in the open market as the Walras model assumes but by negotiations between firm and worker is not novel. For example, in the standard search model discussed in detail in Chap. 3, wages are determined as a Nash equilibrium in the game of how profits are distributed between firm and workers. To be sure, this is different from the model in which wages are determined in such a way that supply and demand are equated in open marketplace. However, the actual determination of wages through labor-management negotiations is not as simple as considering the Nash equilibrium of the game.

Figure 6.9 shows the relationship between the rate of change in wages (wage increases determined in the annual nationwide bargaining called the "spring offence") and the change in corporate profits in the half-century after WWII. If we look roughly at the changes in wages and profits over the past 55 years, there is indeed a positive relationship. From the period of high growth through 2020, both declined together. Formally, as shown in Table 6.1, the correlation between the two is 0.5, although profits fluctuate more than wages. However, this is merely a long-term trend that reflects the fact that the growth rate of the Japanese economy has been declining for the past 55 years after the end of the high growth period. Because the question of how wages are determined is an annual matter, we must look at the short-term relationship. The correlation between the two (1995–2020) over the past 25 years turns out to be –0.02. There is almost no relationship between the two in the short run. As we briefly explain it, this fact shows that the Nash equilibrium wage determination routinely assumed in search theory and other theoretical models does not hold true.

Let us consider a simple Nash equilibrium. Let Y be the (nominal) value added produced. W is the nominal wage, and profit is Π. The firm and the workers (trade union) negotiate in such a way that they maximize profit Π (= Y -- W) and nominal wage W, respectively. The Nash equilibrium is the W that maximizes

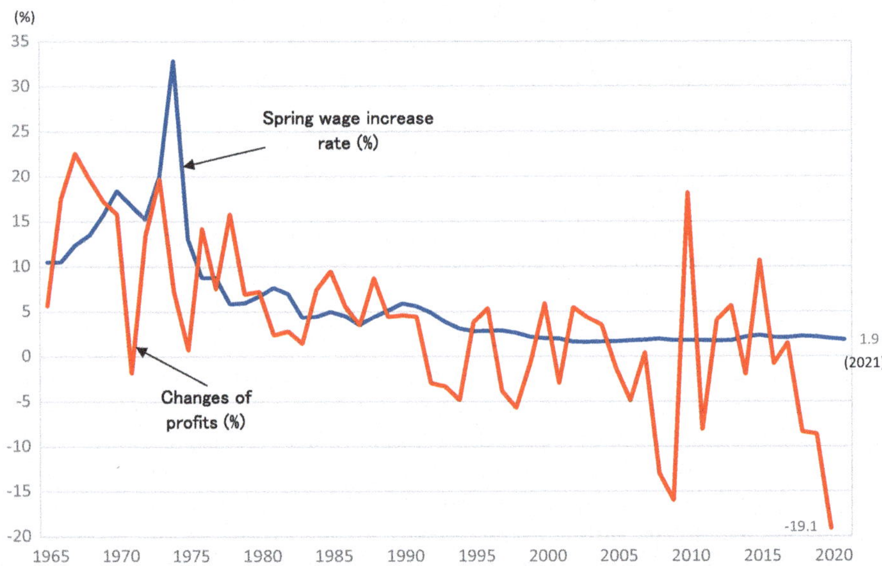

Fig. 6.9 Wages and profits. *Source* Ministry of Health, Labor and Welfare, *Survey on Wage Increase*. Cabinet Office, *SNA (National Accounts of Japan)*. *Note* Both wages and profits are annual rates of change

Table 6.1 Wages and profits

	1965–2020		1995–2020	
	Mean (%)	S. D. (%)	Mean (%)	S. D. (%)
Nominal wages	6.1	6.0	2.1	0.4
Profits	3.8	8.8	−1.0	8.1
Correlation between wages and profits	0.50		−0.02	

Source Ministry of Health, Labor and Welfare, *Survey on Wage Increase*. Cabinet Office, *SNA (National Accounts of Japan)*

$$(Y - W)^{\alpha} W^{1-\alpha} \tag{6.17}$$

Then, we obtain

$$W = \left(\frac{1-\alpha}{\alpha}\right)\Pi \tag{6.18}$$

If (6.18) holds, then the rate of change in wages and profits are equal. However, as we see in Table 6.1, the short-term correlation between the two is almost zero. The same problem can be seen from a different angle. If (6.18) holds, the distribution of labor income is constant at $1-\alpha$, but in reality, it is not constant at all; for Japan, see Yoshikawa (1994). The simple Nash bargaining model does not explain reality well.

We may rewrite the Nash bargaining between the firm and the union as follows.

$$(Y - W)^{\alpha}(W - W^*)^{1-\alpha} \tag{6.19}$$

Firm is interested in profit $\Pi = Y - W$, but workers are not interested in the wage itself, but in $W - W^*$, namely, how much the wage W exceeds the "base" wage level W^*. Here, the "base" wage is a kind of "social norm".

In this case, the wage determined in Nash equilibrium is

$$W = \alpha W^* + (1 - \alpha)Y \tag{6.20}$$

Nominal wages depend not only on the share of the value-added Y produced in this period but also on the "standard" wage level, W^*. In other words, wages are not simply determined by negotiation between labor and management within a firm but depend strongly on the "social norm". What plays an essential role in the determination of W^* is, as Hicks argues, nothing less than *fairness*.

The concept of fairness cannot simply be given a general definition. Obviously, it depends on the historical-social context. Hicks points out the following fact about wages in postWorld War I Britain.

> Then came the war—World War I. By 1920 the wage index (with 1914 = 100) had risen to 280; but by 1923 it had fallen from that to 194. A drop of nearly one-third, and no social revolution! And subsequently, very remarkably, the stickiness was resumed (Hicks, 1974, 67).

As Hicks says, the reason why a more than 30 percent decline of wages in 1920-23 was possible is that prior to that time, workers knew that the wage increase during the War was abnormal. In other words, the fall brought wages back to the normal level. After returning to normal, wages again showed rigidity. The wage we call "normal" here is the wage that is fair and appropriate. Importantly, it is historically formed and is anchored in the past. In fairness, there is no room for expectations about the future.

6.6 Summary

Modern microfounded macroeconomics is in a state of confusion over price and monetary policy with an obsessive focus on expectations. Inflation/deflation is a sustained rise or fall in the general price level. Since the 18th century (or earlier), the quantity theory of money, which holds that the general price level is determined by the quantity of money (money stock), has been a major doctrine. However, Keynes and Schumpeter flatly rejected quantity theory, and even after WWII, until the 1960s, few leading economists supported it. In the UK, Kaldor (1970, 1982) was a harsh critic of monetarism, a modern version of the quantity theory advanced by Milton Friedman. In the United States, too, there were many critics of monetarism, including Samuelson, Tobin, and Solow.

Monetary policy has always been essentially an interest rate policy—except for the "zero interest rate" period in recent years; for Japan, see Yoshikawa (1993). Contrary to Friedman's view, it is a discretionary change in interest rates. Discretionary monetary policy, which is necessary to stabilize the macroeconomy, is a feedback policy of raising or lowering interest rates in response to inflation and unemployment long before the term "Taylor rule" was coined (Yoshikawa, 1981).

However, with the evolution of microfounded macroeconomics over the past 50 years, the quantity theory of money has now become mainstream in the academic community. True, no economist currently vociferously advocates the quantity theory, although Bernanke (2020, 961) notes that the "crude monetarism" was sometime espoused in the early days of quantitative easing, which held that the large increases in monetary base would lead to runaway inflation. Few believe in Friedman/Lucas theory that business cycles are caused by "unanticipated" changes in money stock. However, the idea that prices are determined by money *in the long run* is still a central dogma of the academic community. It is embedded in most microfounded macromodels. In such models, the role of expectations has been misdirectedly emphasized in determining prices/wage. However, apart from oil, raw materials and grains, there is no room for expectations to play a role in the determination of prices of the vast majority of goods and services.

Prices change in such a way as to equate supply of and demand for all the goods and services. Prices serve as a "hand flag signal" to achieve efficient resource allocation in the market economy. This is the message of Walras' general equilibrium theory. However, as we have stated several times, Walras' model is not a legitimate model of the real economy. Most prices and wages are determined in what Hicks calls the fixed price market. In such a market,

> Prices have a social function as well as an allocative function (Hicks, 1974, 85).

Fairness is key to the social functioning of prices. It plays a particularly important role in the determination of wages. We can recall Keynes again in this respect. In *General Theory*, Keynes first paid attention to the fact that nominal wages are determined individually by each firm and sector. Therefore, unlike a cut of real wages due to inflation, a cut of nominal wages cannot bring about a simultaneous and uniform decrease of real wages in the economy as a whole. He went on to explain that workers and unions resist nominal wage cuts because they have a strong interest in their own *relative wages* compared to the wages of other occupations/firms/sectors. Firms understand the workers' resistance. In this way, in a downward rigidity of nominal wages ensues.

Because relative wages are real, not nominal, it is not a money illusion on the part of workers to be concerned with relative wages. Needless to say, relative wages between various occupations, firms, and sectors that workers insist on must be "fair" in some sense. What else can make such a fair relative wage (or wage system) but social forces? Thus, the insistence on relative wages that Keynes cites as an explanation for the downward rigidity of nominal wages is almost synonymous with the fairness emphasized by Hicks and others.

Nevertheless, many economists are bound to reject the introduction of such a concept as "fairness" into economic theory. Nevertheless, whether we call it "fairness", social factors plainly play a major role in the determination of wages. Before we conclude this chapter, it is useful to turn our attention to the long-term evolution of the share of land rent (Fig. 6.10). Land is one of the essential factors of production along with labor and capital. According to neoclassical theory, it must be equal to the marginal products of land. One may recall that the theory of marginal productivity was born in David Ricardo's theory of rent.

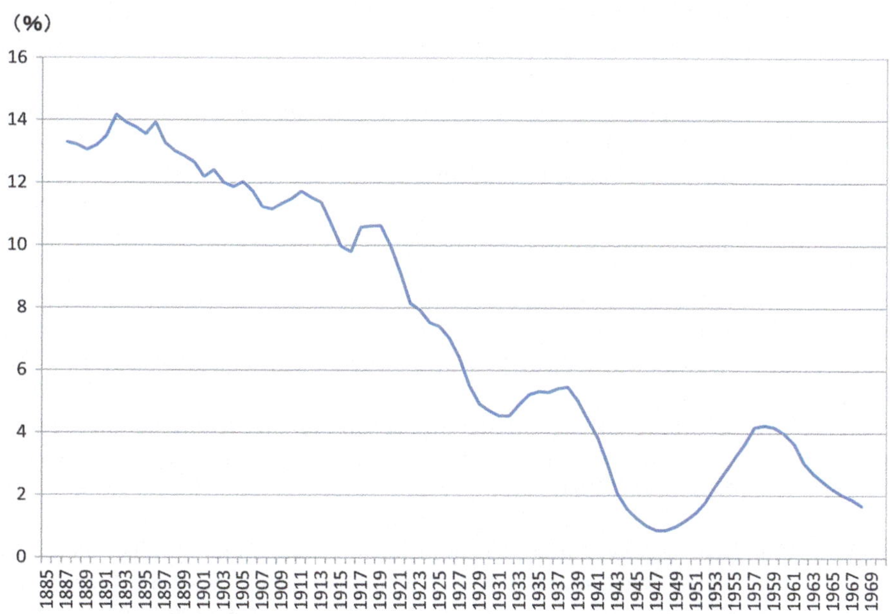

Fig. 6.10 Share of Land Rent. *Note* Figures are the five-year moving averages. *Source* Fukao et al. (2017a, b, 2018)

Now, although the overall downward trend is clear, the share of rent fluctuates rather erratically, so we look at the five-year moving average. During a short period from 1917 to 1927, the share of land rent fell from 10.6 percent to 6.4 percent. It then fell again from 5.5 percent to 0.9 percent for ten years from 1937 to 1947.

According to neoclassical theory, what determines the income share of the production factor is its endowment and technology or the elasticity of substitution of various factors of production. However, it is unreasonable to assume that the production function or technology (elasticity of substitution) and the endowments of labor, capital, and land changed drastically during the two world wars. Rather, the change was brought about socially or politically. To fight the "Total War," it was necessary to suppress the incomes of the landlord class that had earned large "unearned incomes", i.e., land rents. The government fully recognized the fact that soldiers who fought Wars were from working class and peasants, not land owners. In this way, the share of land rents was suppressed by social forces rather than by purely economic forces.

Wages are not the same as rent. However, Keynes pointed out the importance of relative wages, and Hicks emphasized the notion of fairness. The growing inequality that has gained global attention since the end of the twentieth century cannot be fully understood by narrow economic logic alone. Best-selling author Piketty (2013) argues that under capitalism, capital incomes snowballs and inequality grow, but in the United States, where inequality is greatest, it has actually grown within labor income. Stiglitz and others refer to the extraordinary rise in management income as rent-seeking, but this alone is not a sufficient explanation. What allowed this rent-seeking to happen? Arguably, ideologies such as the rise of neoliberalism such as the Chicago School played a major role. It is, after all, a kind of social force.

Of course, wages are not determined by social forces alone. They are also affected by demand and supply in the real economy. This relationship is the Phillips curve (Phillips, 1958, Tobin, 1972). Although it has flattened in many countries, the Phillips curve stands well even in Japan, where nominal wages have been suppressed for many years (Fig. 6.11).

In Japan's Phillips curve shown in Fig. 6.11, there is a truly "abnormal" wage increase in the early 1970s. We must not dismiss this as an outlier. A key to elucidating the mechanism of wage determination is perhaps hidden in the outlier. Let us therefore take a closer look at the movement of nominal real wages in the early 1970s (Table 6.2). Nominal wages increased by 23% and 26% in 1973–74, respectively. The movement of real wages differs considerably depending on whether nominal wages are deflated by the wholesale price index (WPI, currently, corporate price index) or the consumer price index (CPI). Here, we focus on real wages deflated by Consumer Price Index ((5) in Table 6.2). Real wages increased at an annual rate of 3.0 percent in 1974. Plainly, it was very low compared to the average rate of wage increases since the late 1960s, which reached 10%. The drastic change was in response to a fall of the rate of growth of the Japanese economy, but as of the early 1970s, workers were not prepared to fully understand this "structural change", namely, the end of rapid growth. Despite the rapid increase in nominal wages during severe inflation, the rate of increase in real wages (measured in terms of consumer prices) was significantly lower than the level during the period of rapid growth. The experiences of the

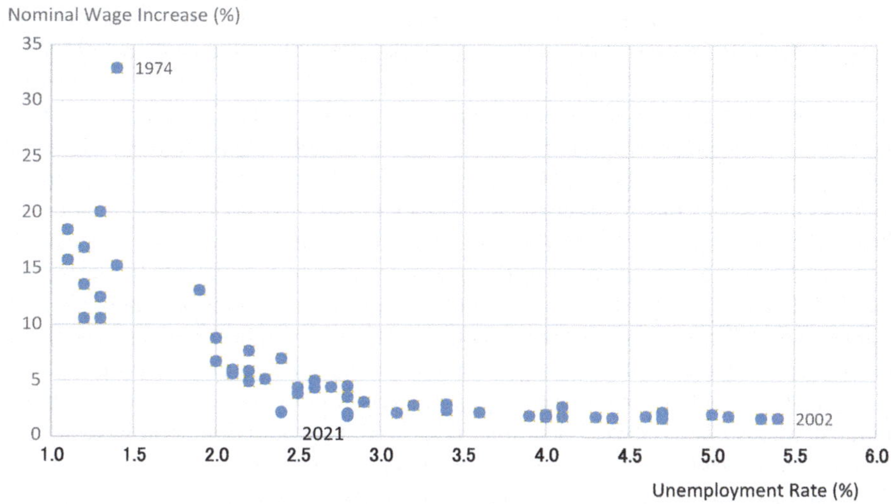

Fig. 6.11 Phillips curve (1965–2021). *Source* Ministry of Health, Labor and Welfare, *Survey on Wage Increase*. Statistics Bureau of Japan, *Labor Force Survey*

early 1970s demonstrate that economic agents—workers and firms—do not share a common macro constraint at all (Fig. 1.1), which in turn shows that the fundamental assumption of modern microfounded macroeconomics does not hold true.

Table 6.2 Wages and prices in the 1970s

	(1) Nominal wages (%)	(2) Wholesale Price Index (WPI) (%)	(3) Consumer Price Index (CPI) (%)	(4) Real Wages (WPI, (1)/(2)) (%)	(5) Real wages (CPI, (1)/(3)) (%)	(6) Working days lost (days)
Ave.67–69	14.8	1.7	4.8	13.1	10.1	2,768
1970	17.5	3.5	7.7	14.0	9.8	3,915
1971	14.0	−1.0	6.3	15.0	7.7	6,029
1972	15.7	1.6	4.9	14.1	10.8	5,147
1973	23.4	15.8	11.7	7.6	11.7	4,604
1974	26.2	27.6	23.2	−1.4	3.0	9,663
1975	11.5	2.7	11.7	8.8	−0.2	8,016
1976	12.3	5.5	9.4	6.8	2.9	3,254

Note Nominal wages are "total monthly wages in manufacturing industry". The wholesale price index corresponds to the current corporate price index compiled by the Bank of Japan. Wage/price figures are all annual rates of change (columns (1)–(5)). The working days lost are labor days lost in dispute. *Source* Ministry of Health, Labor and Welfare, *Monthly Labor Survey, Survey on Labor Disputes*

The worsening of industrial relations between 1974–75 corresponds to these facts.[7] Indeed, in looking at the number of working days lost, the peak in 1974-75 is striking. This is far more than the peak of 6995 days (1948) in the immediate aftermath of the war when labor-management relations were worst. The number of working days lost declined rapidly after 1977, from approximately 500 days to an average of 200 days per year in the late 1980s. 1974–75 was the end of the labor-management conflicts in postwar Japan. Freeman and Rebick (1989) note declining union density during the priod of 1975–87 and demonstrate that much of the decline in density is associated with the inability of Japanese unions to organize new establishments. They attribute this in part to lowered worker interest and stiffened management opposition to unionism following the oil shock, buttressed by unfavorable changes in political and legal environment for collective bargaining and for union organization.

Against this background, in weak labor markets after the 1997–98 financial crisis, nominal wages finally started declining (Fig. 6.5). Low inflation soon gave way to deflation: from 1998 to 2012, Japan's consumer prices fell by a cumulative total of just under 4% with deflation averaging approximately 0.3% per year. Aoyama et al. (2022) shows that the weak bargaining power of workers contributed to the flattening of Japan's Phillips curve. For the U.S., Stansbury and Summers (2020) argue that the declining worker power hypothesis explains sluggish wage growth, declining labor share of income, and low inflation. The experiences during the period of 2013–22 show that unconventional monetary policy could not achieve the inflation target of 2%.

Bernanke (2020) argues that unconventional monetary policy, or rather what he calls the new tools of monetary policy, including quantitative easing (QE), forward guidance and inflation targets, was effective, largely drawing on the U.S. experiences. However, following the Great Recession, the U.S. consumer price, arguably the most important variable of any central bank targets, only slowly recovered the level of 2% in 2017, slid to 1.2% in 2020, and then sharply rose to 8.5% in April 2022. This record is not truly up to Bernanke's claim. In Japan, the two percent target had not been achieved despite the BOJ's nine-year long QE. Prices and wages in the actual world are determined in a very different way than the modern microfounded macroeconomics postulates obsessively emphasizing the role of expectations. Aside from the prices of raw materials and foodstuffs, the prices of manufactured goods and services and wages are affected by not only economic forces but also social forces.

[7] 'Trade union power' is said to have some explanatory power in explaining the long-run movement of labor share in the UK (Phelps Brown & Hart, 1952). Kalecki's (1954) theory that the 'mark-up rate' relative to costs determines the labor share also states that the 'mark-up rate' is influenced by the power relationship between labor and management. For the recent U.S. economy, see Stansbury and Summers (2020).

Chapter 7
Financial Markets and the Real Economy

A well-developed financial market contributes to economic development through efficient resource allocation. However, sharp fluctuations in asset prices occasionally have devastating effects on the macroeconomy. Japan's bubble in the late 1980s and its bursting in the early 1990s left a significant negative legacy for the Japanese economy. The subprime loan and securitization boom of the 2000s in the United States and the Great Recession following the collapse of Lehman Brothers in September 2008 are another example.

Unlike the prices of ordinary goods and services, asset prices such as stock and land prices fluctuate violently. In his *General Theory*, Keynes warned of the danger of "bubbles" in the stock market.

> Speculators may do no harm as bubbles on a steady stream of enterprise. But the position is serious when enterprise becomes the bubble on a whirlpool of speculation. When the capital development of a country becomes a by-product of the activities of a casino, the job is likely to be ill-done. The measure of success attained by Wall Street, regarded as an institution of which the proper social purpose is to direct new investment into the most profitable channels in terms of future yield, cannot be claimed as one of the outstanding triumphs of *laissez-faire* capitalism – which is not surprising, if I am right in thinking that the best brains of Wall Street have been in fact directed towards a different object (Keynes 1936, 159).

In this chapter, we consider asset prices and the relationship between financial markets and the real economy. We first briefly review orthodox neoclassical economics, explain the mathematics behind large fluctuations with a focus on the *power distribution*, and discuss bubbles. On this preparation, we argue that there is a fundamental difference between financial markets and the real economy.

H. Yoshikawa, *Reconstruction of Macroeconomics: Methods of Statistical Physics, and Keynes' Principle of Effective Demand*, Advances in Japanese Business and Economics 3, https://doi.org/10.1007/978-981-19-5264-7_7

7.1 The Neoclassical Approach

Stock prices depend on the real economy. Their 'correct' prices or the fundamental values are the discounted present values of a stream of future dividends/profits. Since business activities, in particular profits, are significantly affected by the state of the real economy, stock prices are also affected by the real economy. More precisely, in the standard neoclassical theory, stock prices are simultaneously determined together with all the supplies and demands in general equilibrium (Diamond 1967). Thus, similar to production and consumption, stock prices ultimately depend on preferences and technologies.

The counterpart of neoclassical economics is the *efficient markets theory* in finance.

> The efficient markets theory reached its height of dominance in academic circles around the 1970s. At that time, the rational expectations revolution in economic theory was in its first blush of enthusiasm, a fresh new idea that occupied the center of attention. The idea that speculative asset prices such as stock prices always incorporate the best information about fundamental values and that prices change only because of good, sensible information meshed very well with theoretical trends of the time. Prominent finance models of the 1970s related speculative asset prices to economic fundamentals, using rational expectations to tie together finance and the entire economy in one elegant theory. For example, Robert Merton published "An Intertemporal Capital Asset Pricing Model" in 1973, which showed how to generalize the capital asset pricing model to a comprehensive intertemporal general equilibrium model (Shiller 2003, 83).

However, there is a long tradition in economics, which questions whether stock prices are truly determined in the way stated above. Many believe that "bubbles" are possible in the market. Whether they are "rational," extraordinary changes in stock prices (either up or down) by themselves may harm the real economy. By any measure, they are not a mere mirror image of the real economy. Throughout history, depressions were often accompanied by falls in stock prices. As early as the 19th century, economists were talking about "financial crises." Kindleberger (1978) in his classics entitled *Manias, Panics, and Crashes* assembles notorious financial crises in world history.

More recently, Minsky (1986) highlighted the possibly disturbing role of stock prices in the macroeconomy and advanced the financial instability hypothesis. It was revived in the 1990s and bore the vast literature. Today, central banks closely monitor asset prices and pay much attention to the stability of the financial system in the conduct of monetary policy.

A crucial problem is whether the stock prices are always equal to their fundamental values. Shiller (1981), in his seminal work, performed ingenious "variance-bound tests" on this issue and drew the following conclusion:

> We have seen that measures of stock price volatility over the past century appear to be far too high — five to thirteen times too high — to be attributed to new information about future real dividends if uncertainty about future dividends is measured by the sample standard deviations of real dividends around their long-run exponential growth path (Shiller 1981, 433).

Naturally, Shiller's seminal work spawned the debate over the alleged excess volatility of stock prices. Rather than accepting that stock prices are too volatile to be consistent with standard theory, a majority of economists have endeavored to reconcile the alleged volatility with the efficiency or "rationality" of the market (Shiller 2003).

One way to explain the volatility of stock prices is to allow significant changes in the discount rate or the required return on stocks, which Shiller (1981) assumes is constant. In fact, in neoclassical macroeconomic theory, the following relationship between the rate of change in consumption, C, and the return on capital, *r,* must hold in equilibrium:

$$-\left[\frac{u''(C)C}{u'(C)}\right]\left(\frac{\dot{C}}{C}\right) = \frac{1}{\eta(C)}\left(\frac{\dot{C}}{C}\right) = r - \delta \tag{7.1}$$

Here, the elasticity of intertemporal substitution η is defined as

$$\frac{1}{\eta(C)} = -\frac{u''(C)C}{u'(C)} > 0 \tag{7.2}$$

In general, η depends on the level of consumption C. Equation (7.1) says that the rate of change in consumption over time is determined by η and the difference between the rate of return on capital, *r,* and the consumer's subjective discount rate, δ. This Euler equation is derived as the necessary condition of the representative consumer's maximization of the Ramsey utility sum. Thus, according to neoclassical macroeconomics, the return on stocks must be consistent with the rate of change in consumption over time in such a way that Eq. (7.1) holds. Therefore, if the rate of return of a stock r is volatile, then consumption C must also be volatile. However, it is well known that the rate of change in consumption ($\dot{C}/C$) is much smaller than that of capital investment (Fig. 7.1).

The results of the tests of Shiller (1981) imply that the volatility of stock prices must come from the volatility of the discount rate or the return on capital r, rather than that of dividends. Since consumption C is not volatile, much less volatile than dividends or profits, it, in turn, suggests that given Eq. (7.1), the volatility of stock prices must be explained ultimately by sizable fluctuations of the elasticity of intertemporal substitution η, which depends on consumption. Consequently, on the representative agent assumption, researchers focus on the "shape" of the utility function in accounting for the volatility of stock prices (Grossman and Shiller 1981). It is not an easy task, however, to reconcile the theory with the observed data if we make a simple assumption for the elasticity of intertemporal substitution, η.

A slightly different assumption favored by theorists in this game is that the utility, and therefore, this elasticity η, depends not on the current level of consumption C_t but on its deviation from the "habit" level, $\widehat{C}_t$, namely, $C_t - \widehat{C}_t$. By assumption, the habit $\widehat{C}_t$ changes much more slowly than consumption C_t itself so that at each

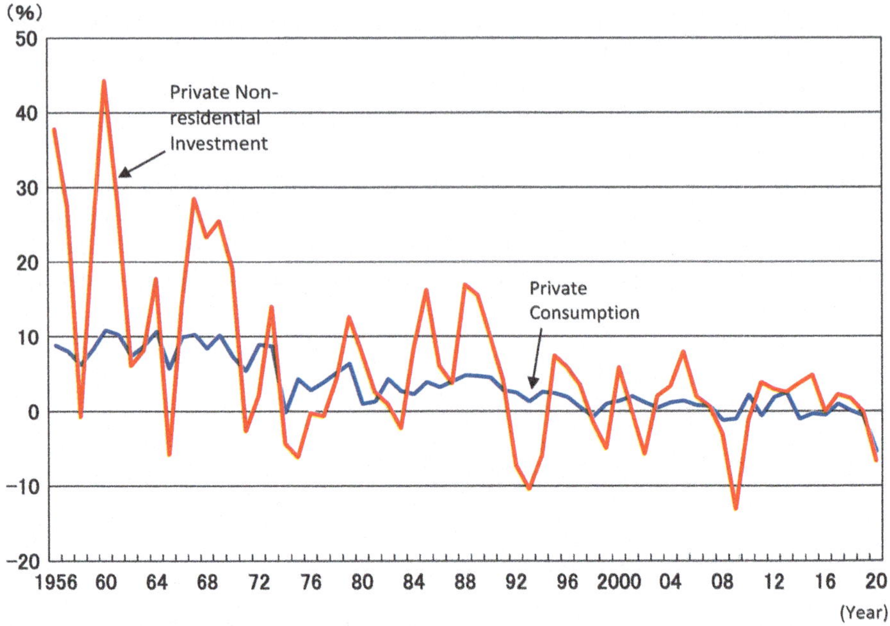

Fig. 7.1 Changes in Consumption and Investment. *Source* Cabinet Office, *SNA (National Accounts of Japan)*

moment in time, $\widehat{C}_t$ is almost constant. The trick of this alternative assumption is that although C_t does not fall close to zero, $C_t - \widehat{C}_t$ can do so to make the elasticity of intertemporal substitution η, now redefined as

$$\frac{1}{\eta} = -\frac{u''\left(C - \hat{C}\right)\left(C - \hat{C}\right)}{u'\left(C - \hat{C}\right)} > 0 \tag{7.3}$$

quite volatile. Campbell and Cochrane (1999) is a primary example of such an approach. Although ingenious, the assumption is not entirely persuasive. Why does the consumer's utility become minimal when the level of consumption is equal to the habit level even if it is extremely high? In any case, this is the kind of end point we are led to as long as we keep the representative agent assumption in accounting for the volatility of stock prices.

Meanwhile, Mehra and Prescott (1985), using a representative agent model, presented another problem for asset prices. They considered a simple stochastic Arrow-Debreu model. The model has two assets, one the equity share for which dividends are stochastic and the other the riskless security. Again, on the representative agent assumption, the "shape" of the utility function and the volatility of consumption play the central role for prices of or returns on two assets. For reasonable values of η, which may be more appropriately called relative risk aversion in this stochastic model, and the U.S. historical standard deviation of consumption growth, Mehra and Prescott calculated the theoretical values of the returns on two assets. The risk premium, namely, the difference between the return on the equity share and the return on the riskless security implied by their model, turns out to be merely 0.4%. However, the actual risk premium for the U.S. stock (the Standard and Poor 500 Index, 1889–1978) against short-term security, such as the Treasure Bill, is 6%. Thus, the standard model with the representative consumer cannot account for such a high risk premium that is actually observed. Mehra and Prescott posed this result as a puzzle. Since then, a number of authors have attempted to explain this puzzle: see Campbell and Cochrane (1999).

The "puzzles" we have seen are, of course, conditional on the assumption of representative agents. Indeed, Deaton (1992) laughs away the so-called "puzzles" as follows:

> There is something seriously amiss with the model, sufficiently so that it is quite unsafe to make any inference about intertemporal substitution from representative agent models...... The main puzzle is not why these representative agent models do not account for the evidence, but why anyone ever thought that they might, given the absurdity of the aggregation assumptions that they require. While not all of the data can necessarily be reconciled with the microeconomic theory, many of the puzzles evaporate once the representative agent is discarded (Deaton 1992, 67, 70).

We second Deaton's criticism. Having said that, here, we note that the standard analyses all focus on the variance or the second moment of asset prices or returns; see Cacchetti et al. (2000), for example, and the literature cited therein. As we will see shortly thereafter, a number of empirical studies actually demonstrate that the variance or standard deviation may not be a good measure of risk. In the next section, we review the mathematics behind the large fluctuations of asset prices.

7.2 Mathematics of Asset Price Fluctuations

Asset prices sometimes fluctuate significantly. In this section, we review the mathematical mechanisms that produce large fluctuations.

Power Distribution

Assuming a normal distribution (Gaussian) is the most standard approach to analyzing stochastic variables. The *central limit theorem* provides the basis. However, at the beginning of the 1960s, Mandelbrot, the father of fractal geometry, found that changes in asset prices cannot be explained by a model that assumes a normal distribution (Mandelbrot 1963a; Mandelbrot and Hudson 2004).

The distribution found by Mandelbrot for changes in asset prices (specifically, the price of cotton on the New York market) was the power distribution that had been found by Pareto (1896) for the income distribution of high-income earners at the end of the 19th century. The probability density function $f(x)$ of a random variable x that follows a power distribution is[1]

$$f(x) = \alpha\left(\frac{1}{x^{\alpha+1}}\right) \quad (\alpha > 0) \quad for\ x \geq 1 \tag{7.4}$$

The probability of an "abnormal" event such as a *black swan* with large x becomes small even under the power distribution. However, compared to the probability of a large x measured by $\exp(-\alpha x^2)$ for the normal distribution, it is much greater in the case of the power distribution. Because the distribution of height is normal, there is no one with a height of 5 m, but if it were a power distribution, there would be a person with a height of 5 m! This fact, which characterizes the power distribution, is sometimes referred to as "fat tail(s)" in the sense that the hem of the probability distribution corresponding to large x becomes thicker.

Under the power distribution, the m-th order moments are

$$\int_1^\infty x^m \alpha\left(\frac{1}{x^{\alpha+1}}\right)dx \tag{7.5}$$

Therefore, there exist no moments of order greater than $m \geq \alpha$. For example, if $\alpha = 1$, there are no first-order moments or even the mean. If $\alpha = 2$, the mean exists, but the second-order moment, or variance, does not. In short, a random variable following a power distribution is a "runaway horse" that is much more violent than that of a normal distribution. In some cases, there may not be even variance.

Mandelbrot (1963), who discovered the power distribution half a century ago, experienced the sorrow of the pioneers. However, it is now common knowledge that fluctuations in stock prices and stock returns cannot be captured by simple normal distribution, and the world of finance has come to use time-series models such as ARCH or GARCH (a generalization of ARCH) in which the variance changes with time. It was physicists and practitioners who called themselves econophysicists who revived the power distribution. A number of empirical analyses showed that fluctuations in stock prices and exchange rates follow the power distribution (Montegna

[1] For the probability density function to be meaningful, $x \neq 0$ so that we have assumed that $x \geq 1$. This assumption is for an explanatory convenience, and not essential. For $f(x)$, a probability density function, we make the power with $\alpha + 1$, not α. Thus, the power of the integrated probability distribution function is α. The power α of distribution function is often called the order of the power distribution.

and Stanley 2000; Stanley et al. 2006).[2] Whether or not second-order moments (variances) exist, i.e., whether or not $\alpha > 2$ is a subtle issue. Mandelbrot (1963a, b) emphasized that there is no variance in the distribution of asset price changes. However, Fama (1965) demonstrated that there exists variance. Mantegna and Stanley (2000, Ch. 9) recapitulates the fact that there exists variance, although it varies over time.

One of the implications of the power distribution is that it is meaningless to discuss the properties of the probability distribution based on low-order moments alone, even if there exists a second-order moment. The consumption-based CAPM and the "equity premium puzzle" by Mehra/Prescott are theoretical analyses focusing on first- and second-order moments, and therefore, it does not make much sense from a starting point.

The question that needs to be asked is what kind of mechanism generates volatile asset price fluctuations that follow the power distribution. In the following, we review the mathematical mechanism that generates the power distribution.

Table 7.1 shows a list of famous studies collected by Newman (2005). The distribution of word frequencies at the top is based on *Moby Dick* by Melville, but the power distribution is not only found for *Moby Dick* [the interested reader is referred to Zipf (1949)]. What is in question is why the power distribution is so universal, or the mathematics of the power distribution.

Limit Distribution of the Sum of Independent Random Variables

Consider the distribution of the sum of independent random variables. The "queen" of the probability distribution is said to be a normal (Gaussian) distribution. The basis for making the normal distribution normal is, as is well known, the central limit theorem.[3]

Central limit theorem

$X_1, \ X_2, \ \cdots$ are independent and identically distributed (*iid*) random variables, with expectation $m = E(X_i)$ and variance $\sigma = V(X_i)$ and third-order moments $E[|X_i|^3]$. Then, converges to a standard normal distribution $N(0, \ 1)$ with $n \to \infty$.

$$S_n = \frac{(X_1 + X_2 + \cdots + X_n) - nm}{\sqrt{n\sigma}}$$

[2] However, there is some dissent in the area of finance. When Mandelbrot died in October 2010, in the *Financial Times* "Mandelbrot's inconvenient truth " was published (Tuesday, October 19, 2010, The Lex Column column). The article introduced Mandelbrot's discovery of the power distribution, argued that this was a rejection of the "efficient market theory" of finance, and concluded that in light of the financial crisis of 2008–2009, Mandelbrot alone should have won the Nobel Prize in economics. Soon after, on October 21, 2010, the *Financial Times* published an article entitled "How Mandelbrot caused confusion", which argued that although the dispersion of stock returns is not constant, the returns can be approximated by a normal distribution by normalizing appropriately with a variable variance, and that Mandelbrot's claims to the theory of finance have only caused unnecessary confusion.

[3] Beginning with Bornoulli, the history of the central limit theorem by de Moivre, Laplace, Gauss, Chebyshev and others is given in the Appendix "History of Probability Theory" of Gnedenko (1997).

Table 7.1 Example of the power distribution

	Quantity	Minimum	Exponent
		x_{min}	α
(a)	Frequency of use of words	1	2.20(1)
(b)	Number of citations to papers	100	3.04(2)
(c)	Number of hits on web sites	1	2.40(1)
(d)	Copies of books sold in the US	2,000,000	3.51(16)
(e)	Telephone calls received	10	2.22(1)
(f)	Magnitude of earthquakes	3.8	3.04(4)
(g)	Diameter of moon craters	0.01	3.14(5)
(h)	Intensity of solar flares	200	1.83(2)
(i)	Intensity of wars	3	1.80(9)
(j)	Net worth of Americans	\$600 m	2.09(4)
(k)	Frequency of family names	10,000	1.94(1)
(l)	Population of US cities	40,000	2.30(5)

Note The details of each distribution are explained in the same paper
Source Newman (2005), 330, Table 1

The central limit theorem states that the sum of independent and identically distributed random variables converges to a normal distribution under the condition that there are low-order moments in the original distribution. The power of this theorem lies in the fact that the original distribution of the random variable X_i that makes up the sum can be any distribution. Probabilistic events occurring in nature and society can be reasonably considered to be the result of the accumulation of numerous disturbances, but the probability distribution of the original stochastic variables comprising the sum is unknown. The central limit theorem is valid regardless of the probability distribution of the original variables. Indeed, there are many phenomena in nature and society that are characterized by a normal distribution, such as height. Thus, the normal distribution has been used as the "queen" of probability distributions in many empirical studies regardless of the field. Many statistical and econometric tests and theorems assume a normal distribution.

For the random variable Y_n, which is expressed as a product (multiplicative process) rather than the sum of a sequence of random variables X_i

$$Y_n = X_1 X_2 \cdots X_n \tag{7.6}$$

If we take the logarithm of the logarithm, $\log Y_n$

$$\log Y_n \equiv S_n \equiv \log X_1 + \cdots + \log X_n \tag{7.7}$$

Thus, the central limit theorem holds for log Y_n under the appropriate assumptions. In this case, Y_n converges to the lognormal distribution.

On the theoretical basis of the central limit theorem, the normal distribution, including the log-normal distribution, is taken as the universal distribution. However, the fact that there are many phenomena to which a power distribution applies instead of a normal distribution has been known since long before Mandelbrot. Pareto's (1896) discovery of the power distribution on income distribution for high-income groups is recognized as the first empirical power distribution ever discovered. In addition, experts found that the power distribution rather than the normal distribution applies to natural disasters such as earthquakes.

Why is the power distribution found so universally? The central limit theorem and stochastic processes had already been elucidated by Levy, Khinchin, Gnedenko, Kolmogorov, Kiyoshi Ito and other mathematicians in the prewar and early postwar periods. However, such findings have not been widely shared by economics and other applied fields.

The problem is to what kind of probability distribution the (suitably normalized) sum $S_n = X_1 + X_2 + \cdots + X_n$ of the stochastic variables $X_1\ X_2\ \cdots\ X_n$ converges when $n \to \infty$. If X_i is *iid*, and there are second (and third) order moments, then the standard central limit theorem leads us to a normal distribution. However, this is not the only possibility. Naturally, the central limit theorem cannot be established unless the assumptions required by the theorem are satisfied. The limit theorems that hold under different conditions than the central limit theorem have been studied in various ways.

For example, if we take a time series of the rate of change in stock prices, the variance is not invariant with respect to time, although at different points in time, it is nearly independent. This corresponds to the case where the random variables X_i are independent of each other but do not follow the same distribution. When the sequence of random variables X_i follows an independent but not identical distribution, the sum $S_n = X_1 + X_2 + \cdots + X_n$ converges to an *infinitely divisible probability distribution* under certain conditions (Khinchin's limit theorem).[4]

The infinitely divisible distribution appearing in Khinchin's limit theorem is defined as follows: for any natural number *n*, there exists a distribution μ_n, and the distribution μ is defined by the n convolution of μ_n, μ_n^n

$$\mu = \mu_n^n \tag{7.8}$$

[4] For this theorem to hold, no single independent random variable X_i has a dominant effect on the partial sum S_n; in other words, as n grows larger, the effects of all the individual X_i's must be negligible. Chapter 4 of Gnedenko and Kolmogorov (1954) explains the general limit theorem.

If then, μ is said to be infinitely decomposable.[5] Because the convolution corresponds to the multiplication of the characteristic function, if the characteristic function (Fourier transform) of the probability distribution μ is $\hat{\mu}$, then there is a characteristic function equal to the n-squared root of $\hat{\mu}$ if μ is infinitely divisible. For infinitely divisible distributions, the general form of the characteristic function was obtained by Levy and Khinchin (Gnedenko & Kolmogorov 1954, 76, Theorem 1). The familiar textbook distributions, such as normal, Cauchy, Poisson, negative binomial, exponential and gamma distributions, are all infinitely divisible, although uniform and binomial distributions are not infinitely divisible. Khinchin's limit theorem suggests that the most important family of probability distributions for us is the infinitely divisible distribution.

The power distribution is also included in the family of infinitely divisible distributions. Note that this family of infinitely divisible distributions includes both distributions with moments, especially second-order moments, and distributions without them. We have already seen that in the case of the power distribution, the second-order moment may not exist depending on the degree. It is well known that there is no variance in the Cauchy distribution, which is also infinitely divisible.

Now let us consider the sum $S_n = X_1 + X_2 + \cdots + X_n$ of the *iid* stochastic variables X_i. That is, we again consider a narrower range than Khinchin's limit theorem. However, it is not assumed that low-order moments exist in the original distribution. Thus, the standard central limit theorem may not be valid.

For any natural number *n, if* the probability distribution $\mu(S_n)$ of S_n is the same as the distribution $\mu(X_i)$ of X_i, we call this distribution a *stable distribution*. Although the general form of the characteristic function of the stable distribution has been clarified by Levy and Khinchin, the exact form of the distribution is not always clear. Nevertheless, we know several important things about stable distributions. The normal and Cauchy distributions are stable distributions. For a stable distribution $P(x)$ other than the normal distribution, which is symmetric on both sides of the mean in the range of a large absolute value of x, the following approximation holds.

$$P(x) \sim \frac{1}{|x|^{\alpha+1}} \qquad (0 < \alpha < 2) \quad \text{for large } |x| \tag{7.9}$$

In other words, as the absolute value of x increases, it approaches the power distribution of order α. Here, α characterizes the stable distribution ($0 < \alpha \leq 2$), and $\alpha = 2$ corresponds to the normal distribution. In (7.9), α is less than 2, so there are no second-order moments. In fact, second-order moments exist only in the normal distribution (in the case of $\alpha = 2$) (Samorodnitsky and Taqqu 1994, 18). A family of a stable distribution is a subset of an infinitely divisible distribution. Therefore, the relationship between them can be expressed as shown in Fig. 7.2.

[5] μ does not have to be one-dimensional and can generally be multidimensional, but we do not specify the dimension of μ because it is not particularly necessary here. See Feller (1971) and Sato (1990) for infinitely divisible distributions.

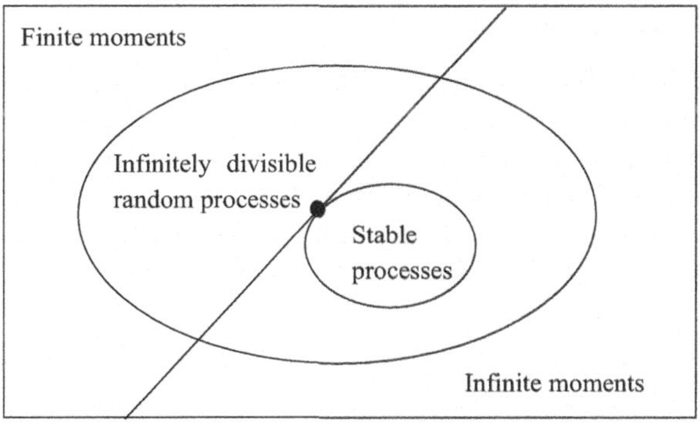

Fig. 7.2 Infinitely Divisible Distribution and Stable Distribution. *Source* Mantegna and Stanley (2000, Fig. 4.3). *Note* The bold dot in the center indicates the normal distributions.

If some phenomenon under analysis appears as a result of the sum S_n of many random variables X_i, the distribution we should expect to see is an infinitely divisible distribution from Khinchin's limit theorem unless the distribution of X_i is identical. The power distribution is indeed included in the family of infinitely divisible distributions. However, there are many other infinitely divisible distributions in addition to the power distribution.

$$f(x) = \alpha\left(\frac{1}{x^{(\alpha+1)}}\right) \quad (\alpha > 0) \tag{7.10}$$

On the other hand, if there is no significant difference between the distributions of X_i and the sum S_n, the family of stable distributions, which is a subset of infinitely divisible distributions, is a good candidate. In this case, the existence of the second-order moment is an important point. If the second-order moment (i.e., variance) exists, the distribution is normal, whereas if they do not exist, it is a power distribution of less than two degrees (especially for large x). We may note that in the space of a real number with parameter α $(0 < \alpha \leq 2)$, which characterizes the stable distribution, the normal distribution has Lebesgue measure zero because it is the case when $\alpha = 2$. On the other hand, the exponent of the power distribution is $0 < \alpha < 2$ and, therefore, has a measure of 2.

From the limit theorem of the sum of independent random variables, what distribution should we expect to see in natural and social phenomena? It is no longer justified to focus only on the normal distribution based on the standard central limit theorem. According to Khinchin's limit theorem, which removes the assumption that the distributions of the original random variables constituting the sum are identical, the *infinitely divisible distribution* is the set of strong candidates. Needless to say, the normal distribution is one of the most important distributions among them. That is why we can find a normal distribution in many natural-social phenomena. However, the generalized limit theorem suggests that it is quite natural for the power distribution to be found, particularly for the tail of the distribution where the absolute value of a random variable is large, if not in the whole distribution.

Entropy Maximization

It is useful to summarize the problem in terms of entropy maximization. Entropy maximization [Eq. 2.13 in Chap. 2] is the most fundamental concept in statistical physics, and we explained in Chap. 3 how it is applied to the microfoundation of Keynesian economics. Here, we consider the relationship between entropy maximization and the resulting probability distribution under different constraints.

For a continuous variable x, the entropy H is

$$H = -\int P(x) \log P(x) dx \tag{7.11}$$

(1) Let the range of x be $(0, \infty)$, and the mean (first-order moment) shall be given.

$$\langle x \rangle \equiv \int_0^{\infty} x P(x) dx = \overline{\mu} \tag{7.12}$$

The distribution $P(x)$ that maximizes the entropy H under constraint condition (7.12) is an *exponential distribution*, i.e., Boltzmann distribution. Needless to say, it is this case that we considered in Chap. 3. The effective demand constraint (3.32) in Chap. 3 is nothing but the condition for a constant mean, namely, (7.12).

(2) Let x be $(-\infty, \infty)$, and the second-order moment is assumed to be given.

$$\langle x^2 \rangle \equiv \int_{-\infty}^{\infty} x^2 P(x) dx = \overline{\sigma}^2 \tag{7.13}$$

The distribution $P(x)$ that maximizes entropy H under the constraint (7.13) equation is a *normal distribution*.

$$P(x) = \frac{1}{\sqrt{2\pi}\sigma} \exp\left(-\frac{x^2}{2\sigma^2}\right) \tag{7.14}$$

In this case, entropy H is given by

$$H = \log \sigma (2\pi e)^{\frac{1}{2}} \tag{7.15}$$

Thus, the normal distribution is obtained from entropy maximization under the constraint of constant second-order moments.

(3) Finally, the mean of $x = \log X$ of the random variable X ($X \in (0, \infty)$)

$$\langle \log X \rangle = \langle x \rangle = \int_0^\infty x P(x) dx = \overline{\mu} \tag{7.16}$$

is assumed to be constant.

Consider the entropy H maximization with respect to $P(x)$. $P(x)$ is the distribution of $x = \log X$. Entropy H is also defined with respect to the logarithm $x = \log X$, not X. Thus, on a symbolic level, we can keep Eq. (7.11) as it is. The problem of maximizing Eq. (7.11) under the constraint Eq. (7.16) is the same as in (1), so the distribution of x, $P(x)$, becomes an exponential distribution (Boltzmann distribution), namely,

$$P(x) = Ce^{-\alpha x} \qquad (\alpha > 0,\ C > 0) \tag{7.17}$$

Because the random variable X is $X = e^x$, the distribution $\hat{P}(X)$ of X is

$$\hat{P}(X) = e^{-x} P(x) = e^{-x} Ce^{-\alpha x} = C(e^x)^{-(\alpha+1)} = CX^{-(\alpha+1)} \tag{7.18}$$

It is a "power distribution". Thus, the power distribution is derived from entropy maximization under the condition of the mean constant logarithmic value of the random variable X.[6]

We can maximize entropy H under different constraints (first-order and second-order moments are constant) and obtain exponential distribution, normal distribution, and power distribution, respectively.

Entropy H in (7.11) is Shannon's entropy. Shannon entropy was recently generalized to Tsallis entropy.

$$T(q) = \frac{1 - \sum_{i=1}^{k} P_i^q}{q - 1} \tag{7.19}$$

The relationship between Tsallis entropy and power's law has attracted much attention. Tsallis entropy (7.19) depends on parameter q, and when $q \to 1$, Shannon entropy H *is* obtained. See Suzari (2010) for the relationship between Tsallis entropy and power distribution.

Multiplicative Process

Among the skewed distributions with *fat tails*, the lognormal distribution is the most famous. As we saw above, the log-normal distribution is obtained by applying the central limit theorem for a random variable expressed as the product of random variables X_i, which are independent and follow the same distribution (taking the

[6] Levy and Solomon (1996) show this in a different way. As the title of their paper states, the "Power distribution" is a "logarithmic version of the Boltzmann distribution".

logarithm). The following theorem by Kesten (1973) and Goldie (1991) is known about generalized stochastic processes in the multiplicative process.

We define a generalized multiplicative process as follows. The random variable Y_t is defined as

$$Y_t = M_t Y_{t-1} + Q_t \qquad (t = 1,\ 2,\ \ldots) \tag{7.20}$$

where M_t and Q_t are a pair of *iid* random variables.

Theorem (Kesten/Goldie)

For the multiplicative process, or stochastic difference Eq. (7.20), if

$$E(\log|M_t|) < 0 \qquad (t = 1,\ 2,\ \ldots) \tag{7.21}$$

is satisfied, the distribution of Y_t converges to a single probability distribution. In addition to the above condition, if $Q/(1-M_t)$ is not a constant multiple of $(1-M_t)$ (nondegenerate), and furthermore,

$$0 < E(|Q_t|^\alpha) < \infty \qquad (\alpha > 0) \tag{7.22}$$

$$E(|M_t|^\alpha) = 1 \tag{7.23}$$

$$E(|M_t|^\alpha \log+|M_t|) < \infty \qquad (\log+x = \max(0,\ \log x)) \tag{7.24}$$

hold, and there exists a positive number α satisfying the above, then the tail of the limit distribution $(t \to \infty)$ of Y_t follows the power distribution with exponent α. That is,

$$P(|Y_t| > x) \sim x^{-\alpha} \tag{7.25}$$

holds.

From this theorem, we can see that there is a close relationship between the power distribution and the multiplicative process. Roughly, the multiplicative process is a hotbed of the power distribution. The order α of the power distribution is obtained from the condition (7.23) that the expected value of the α order moment of the growth factor M_t is 1.

Lognormal Distribution

For the multiplicative process, the logarithmic normal distribution can be obtained by taking the logarithm and applying the central limit theorem as the limit distribution. Therefore, there are still some who argue that the multiplicative process is not essentially different from the additive process. Redner (1990) explains that this is wrong using a binomial distribution.

Consider the following multiplicative process based on binomial distribution. Starting from the initial value 1, multiply Z_1 with probability p and Z_2 ($Z_1 > Z_2 > 0$) with probability q N times, and denote the result by P. Let $\langle P \rangle$ be the mean of P, and let $\hat{P}$ be the value of P that is most likely to occur (i.e., the value that has the greatest probability, or mode).

$$\langle P \rangle = \sum_{n=0}^{N} {}_N C_n p^n q^{N-n} Z_1^n Z_2^{N-n} = \sum_{n=0}^{N} {}_N C_n (pZ_1)^n (qZ_2)^{N-n} = (pZ_1 + qZ_2)^N \tag{7.26}$$

In contrast, $\hat{P}$ maximizes ${}_N C_n p^n q^{N-n}$. Therefore, we obtain

$$\hat{P} = (Z_1^p \, Z_2^q)^N \tag{7.27}$$

Note that in reality, we only have a finite sample and, therefore, that P is expected to be close to $\hat{P}$ rather than $\langle P \rangle$. For the Additive Process, the difference between $\hat{P}$ and $\langle P \rangle$ can be considered to be small. However, in the multiplicative process, the two diverge greatly. As shown by Redner (1990), in the case of binomial distribution, for $1/3 < p < 1/2$, $\langle P \rangle$ diverges as $N \to \infty$ while $\hat{P}$ converges to zero!

In the multiplicative process, even though the probability is small, extreme events have great impacts on the mean (more generally moments). In the case of the multiplicative process based on binomial distribution, the log-normal distribution gives a correct approximation about the moments only when the probabilities of Z_1 and Z_2 are not significantly different ($p \cong q$) or when Z_1 and Z_2 are almost equal ($Z_1 \cong Z_2$).[7] In short, in general, we cannot know exactly the probabilistic nature of the multiplicative process using the lognormal distribution.

Other Mechanisms

There are various other mechanisms that produce the power distribution in addition to those explained above (see Newman 2005, Sect. 4, 336–348).

(1) *Yule Process*: In 1922, Willis studied the classification of organisms and found that the distribution of the number of species within a genus follows the power distribution. The graph with the logarithm of the number of species in a genus on the horizontal axis and the logarithm of the number of genera (cumulative) on the vertical axis formed a beautiful straight line. This linear graph is said to be the pioneering work in representing the power distribution as a linear graph on both logarithmic axes.

[7] Redner (1990, 270) points out that the application of the log-normal distribution is based on a misuse of "Laplace's method" in calculating the integration of moments.

Yule proposed the following stochastic process to explain Willis' findings. Suppose there are n genera that each contain a different number of species. Occasionally, a new genus is born stochastically. The number of species in each genus depends on the number of species already in the genus, and the greater the number of species, the greater the probability that a new species will join the genus. In this model, the number of genera n increases with time, and the asymptotic ($n \to \infty$) distribution of the fraction p_k of a genus with k number of species is a *beta distribution* (see Newman (2005, 340–341) for details).

$$p_k = \left(1 + \frac{1}{m}\right) B\left(k,\ 2 + \frac{1}{m}\right) \tag{7.28}$$

The tail of the beta distribution $B\ (a, b)$ is

$$B(a,\ b) \sim a^{-b} \tag{7.29}$$

Thus, we know that the tail *of* p_k follows the power distribution,

$$p_k \sim k^{-\alpha}, \qquad \alpha = 2 + \frac{1}{m} \tag{7.30}$$

The Yule Process is a model of species differentiation, but it is a general model with a very wide range of applications, as seen, for example, by interpreting the genera as cities and the number of species as the population of each city. The two-parameter Poisson/Dirichlet distribution is a model similar to the Yule Process.

(2) *Lévy Flight*: It is well known that random walk is an important stochastic process with a very wide range of applications, and its limit is a normal distribution. However, the standard random walk assumption that the flight width of a leap is constant in each period is very restrictive. This should be immediately clear if we consider, for example, a grasshopper leaping in a field.

Lévy proposed the following model as an alternative to the standard random walk model. For simplicity, the dimension is assumed to be one. The grasshopper leaps left and right along a one-dimensional line, but the width of the leap is not constant but obeys the following distribution:

$$\begin{aligned}&\pm a \text{ with probability } c/2\\&\pm \lambda a \text{ with probability } c(1-c)/2\\&\pm \lambda^j a \text{ with probability } c(1-c)^j/2\end{aligned} \tag{7.31}$$

where $a > 0$, $1 > c > 0$, and $\lambda > 1$.

In other words, the grasshopper can jump infinitely far to the left and right with the minimum unit of a. The larger the flight range is, the smaller the probability of such a leap becomes. This generalized random walk is sometimes called "Lévy flight".

The case of $c = 1$ in (7.31) corresponds to the standard random walk. As one can easily imagine, Lévy flight is "wild" compared to the standard random walk, which has a constant flight span. The standard random walk yields a normal distribution, whereas Lévy Flight yields a power distribution.

Although the probability is very small, it is perhaps unrealistic for a grasshopper to fly to the opposite side of the earth. Therefore, we may assume that the probability is zero when the absolute value of one leap x exceeds a certain value $|x| > m$ and consider the distribution as shown in (7.31) within the range of $|x| \leq m$.

$$0 \text{ for } x > m > 0$$

$$P(x) = cP_L(x) \text{ for } -m \leq x \leq m$$

$$0 \text{ for } x < -m$$

where $P_L(x)$ is the symmetric Lévy flight explained above. This is the *truncated Lévy flight,* which we will use later in Sect. 7.4.

Unlike the Lévy flight in which the length of a jump is unbounded, the truncated Lévy flight has a limit ($m > 0$) on the length of a jump. Because the truncated Lévy flight has a finite variance, the probability distribution of the sum of n random variables from this distribution, $P(S_n)$, converges to a normal distribution (the central limit theorem). The question is how quickly $P(S_n)$ will converge. When n is small, the Lévy flight well approximates $P(S_n)$. Thus, there exists a crossover value of n, n^* such that

$$\text{For } n << n^*, \quad P(S_n) \sim \text{The Lévy Flight}$$

$$\text{For } n >> n^*, \quad P(S_n) \sim \text{The Normal Distribution}$$

This example illustrates the point that in general, the kind of probability distribution we obtain depends on the sample size, n; see Sect. 8.4 of Mantegna and Stanley (2000) for further details.

(3) *Self-organized* criticality: In statistical physics, it is well known that a power distribution is observed in conjunction with a phase transition (a phenomenon in which the properties of materials such as solids, liquids, gases, and magnetism change qualitatively). However, critical phenomena that generate the phase transition occur only when the parameters controlling the system take a particular value; for example, in the case of water, the critical temperature is either 0 or 100 degrees Celsius. Therefore, as it is, the theory has difficulty explaining that power distribution is universally observed in natural economic and social phenomena.

In other words, it is a weak logic for explaining the power distribution unless we can explain why the control parameters take a particular value.[8]

Self-organized criticality has been devised to solve these problems. It is dynamics that leads to critical phenomena with power distribution regardless of the initial conditions. Bak has a book with the slightly exaggerated title *How Nature Works: The Science of Self-Organized Criticality*.

Bak, Chen, Schenkman and Woodford (1993) show that in an economy consisting of a large number of firms linked in an input–output chain, through the mechanism of self-organized criticality, stochastic fluctuations of demand for final goods lead to a power distribution of macro production. This model can be interpreted as a Keynesian analysis of the "multiplier process". The textbook multiplier analysis makes the assumption that an increase in demand induces the same amount of production. However, depending on the level of inventory stock initially held by the firm, these assumptions may not necessarily be met (Hicks 1974, Chl. 1). Bak *et al.*'s model explicitly analyzes the movement of production in response to the level of inventory and demonstrates how macro fluctuations are generated.

It is not justified to consider the normal distribution as "normal" based on the standard central limit theorem. The normal distribution is certainly one of the most important distributions. However, it is not surprising that the power distribution is found in many natural-economic and social phenomena. We must recognize that it is to be expected based on a casual overview of the mathematics that generates the power distribution.

7.3 The Bubble

Let us now consider the bubble. The "bubble" is defined as the deviation of an asset's price from its fundamental value. In the case of stock prices, the fundamental value is the discounted present price of future earnings (dividends). The greatest problem is that fundamental value is an *ex ante* concept. Regardless of how extraordinarily asset prices changed *ex post*, it may not be a "bubble" from an *ex ante* perspective. This is the consensus view of economists and finance experts. In an interview with *New Yorker* (November 1, 2010), Eugene Fama, winner of the 2013 Nobel Prize in economics for his work on *efficient market theory*, said.

> I don't even know what a bubble means. These words have become popular. I don't think they have any meaning (Cassidy 2010).

For Fama, "bubble" is a completely meaningless concept in analyzing asset price movements.

Meanwhile, a model of "rational bubbles" was devised for the purpose of clearly defining the bubble. Let us take Blanchard's (1979) and Blanchard and Watson's

---

[8] For the same reason, Javanovic's (1987) model, which attempts to explain how the accumulation of micro shocks produces macro variation, also lacks robustness.

(1982) model of the "rational bubble" as an example. In this model, the asset price bubble B_t follows a divergent path with probability π $(0 < \pi < 1)$ every period, while with probability $1 - \pi$, the bubble bursts and B_t falls to zero. To make it possible for the bubble to burst and then re-emerge, B_t has an additive. *iid* disturbance ε_t. Thus, B_t follows

$$B_t = a_t B_{t-1} + \varepsilon_t \tag{7.32}$$

where a_t is.

$$a_t = a > 1 \text{ with probability } \pi$$

$$a_t = 0 \text{ with probability } 1 - \pi \tag{7.33}$$

a_t in (7.33) is also *iid.* Therefore, B_t *is* the multiplicative process defined in Sect. 7.2.

Now for B_t to be a "rational" bubble,

$$B_t = \beta E(B_{t+1}|\Omega_t) \qquad (0 < \beta < 1) \tag{7.34}$$

must hold. Ω_t is the information set for period *t,* and β is the discount factor. From (7.32) and (7.34), we obtain

$$a = \frac{1}{\pi\beta} > 1 \tag{7.35}$$

That is, the rate of progression of the "rational" bubble a must be equal to the inverse of $\pi\beta$.

We apply the Kesten/Goldie theorem described in Sect. 7.2 to this model. Because the condition of the theorem is satisfied, the tail of the limit distribution of B_t becomes a power distribution. Namely, if x is large, then

$$\text{Prob } (B_t > x) \sim x^{-\alpha} \tag{7.36}$$

The order of the power distribution (7.36) αcan be obtained from the condition of Eq. (7.23) of the theorem. Assigning the variables of this model to (7.23), we obtain

$$E(a_t^\alpha) = \pi a^\alpha = 1 \tag{7.37}$$

We delete a by using (7.35) and take the logarithm to obtain

$$\alpha = \frac{\log(1/\pi)}{\log(1/\pi) + \log(1/\beta)} < 1 \tag{7.38}$$

In other words, in the Blanchard/Watson model of "rational" bubbles, the order of the power distribution of stock prices including bubbles is less than one. However, according to most empirical studies, the order of the rate of change in stock prices is approximately three. Therefore, the "rational" bubble model for stock prices is rejected (Lux and Sornette 2002). Conventional economics is often content to create these theoretical models to derive conditions such as (7.35). Using the Kesten/Goldie theorem, we are able to empirically reject this model. Sornette (2003) argues that even the standard stock price model meets the same problem. The "rational bubble" model does not allow for much progress.

Plainly, bubbles are not always present in financial markets but happen only occasionally. Otherwise, the financial market is nothing but a casino and would not justify huge social costs for its maintenance! Indeed, Johansen and Sornette (1998) found that financial crashes are "outliers" in the sense that 1% to 2% of the largest falls of asset prices are not at all explained by the exponential, or its extension the stretched exponential null hypothesis.

Takayasu et al. (2005) also empirically identifies two different phases in the financial market. Specifically, using their own PUCK (Potential of Unbalanced Complex Kinetics) model, they find two different types of potential functions for tick time data on foreign exchanges, one "stable" and the other "unstable". In the former, price tends to return to the "normal" once it deviates, whereas in the latter, it deviates further, producing large changes in price and even crashes. As one might easily imagine, the unstable potential function is generated by the "herd behavior" of market participants (Yamada et al. 2009). Although econophysicists take changes of potential function from stable to unstable one or the other way round as stochastic phenomena, people rarely herd in a purely stochastic manner. Often, there is a major "story" that drives people to herd. Thus, bubbles are full of well-documented episodes (Kindleberger 1978).

An array of studies surveyed by Sornette (2014) indeed advance the broad picture of financial markets in which imitation, herding and positive feedbacks lead *endogenously* to the development of instability. While the efficient market hypothesis provides a useful representation of financial markets *in normal times*, we do observe abnormal regimes where the anchor of the fundamental price becomes shaky, and bubbles/crashes occur. Bubbles are always accompanied by "irrational" market participants; "Rational bubbles" simply do not exist. This is certainly in good accordance with common sense.

It is then a reasonable research agenda to detect the characteristics of such an "abnormal" market in which bubbles might occur. Econophysicists report notable findings in this regard. For example, Sornette (2003) argues that an asset market crash is not exactly a critical or singular point itself, but the triggering rate is strongly influenced by the proximity of the critical point; namely, the closer to the critical point, the more likely a crash occurs. He then puts forward the analytical concept of log-periodicity. This means that the probability or hazard rate of crashes does not monotonously accelerate but often exhibits oscillations with frequencies accelerating as the market approaches the critical point. Sornette (2003) and his associates demonstrate that log-periodicity is, in fact, instrumental not only in the analysis of past crashes but also in the prediction of future crashes. Watanabe, et al. (2009) also demonstrates that their PUCK model can be used in the prediction of future crashes.

Now, it is well recognized that extraordinary booms in stock/real estate markets are very often, if not always, supported by aggressive bank lending. Here comes Minsky (1986), who was never weary of emphasizing the danger of unstable financial markets. The behavior of banks changes over time; see Brunnermeier (2009) for changes in the U.S. banking during the period preceding the 2007–2008 financial crisis. "Bankers can be speculators", Minsky (1975, 121) says. Following Keynes, Minsky (1975) states:

> It is finance that acts as the sometimes damping, sometimes amplifying governor for investment. As a result, finance sets the pace for the economy (Minsky 1975, 130).

In addition to investors participating in asset markets, financial institutions such as banks play an important role in bubbles. Minsky left a kind of paradoxical aphorism, "stability is destabilizing", which means that a stable market eventually changes the behavior of banks and urges them to take high risks.

Ultimately, the fundamental problem with conventional economics is that it is too much concerned with "*ex ante* rationality". It is actually the massive changes in asset prices, the collapse of prices in particular, that cause the major damage to the macroeconomy. Whether price formation is "rational" before the collapse is secondary.

Until 2008, when Lehman Brothers collapsed and the financial crisis and the Great Recession followed, the U.S. Fed had consistently taken the position that bubbles could not be predicted in advance and that the problem could be solved by quickly lowering interest rates when the bubble burst. This is the so-called Fed View. On the other hand, the Bank for International Settlements (BIS) argued for the BIS View, which stated that "the signs of a bubble can be observed, albeit imperfectly, and once the bubble bursts, it will not be easy to solve the problem, even if interest rates are lowered," and sounded a warning about subprime loans in the United States. Today, after the financial crisis and the Great Recession, it is clear that the Fed View was wrong and that the BIS View was right. Behind the advocacy of the Fed View by mainstream macroeconomists was a commitment to the efficiency of financial markets.

7.4 Difference Between Financial Markets and the Real Economy

Asset prices move quite differently from the prices of ordinary goods and services. Many people intuitively feel that asset price fluctuations are far greater than the movements of the real economy. Our wages/salaries rarely double in a year. However, it often happens with stock and land prices. Where do the differences between the real economy and asset markets come from?

In this section, we explore the underlying mechanism that generates power law distributions for returns of financial assets on one side and exponential distributions for real economic variables such as real GDP or consumption on the other (Canning et al. 1998).

As we explained in Sect. 7.2, the truncated Lévy flight, depending on parameters, can generate a wide class of probability distributions, including power laws and exponential distributions. In what follows, we consider a particular model of truncated Lévy flight that nests both power laws and exponential distributions (Aoki and Yoshikawa 2007, Ch.10). The model we consider is an adapted or modified version of Huang and Solomon (2001).

Real Economy

We first consider a model of the real economy. The economy consists of N agents or units. For the sake of expositional convenience, we call the variable of interest "consumption." It may be "production," and in that case, the aggregate variable is GDP. N sectors or units may be interpreted either as N types of consumers or as N types of consumption goods. Interpretation of the model can be very flexible. We will shortly consider stock price using the same model.

The aggregate consumption at time t, $C(t)$, is nothing but the sum of the individual consumptions:

$$C(t) = c_1(t) + \cdots c_N(t) \tag{7.39}$$

Here, $c_i(t)$ is the consumption of the type i good or the i-th consumer's consumption. The argument t stands for calendar or real time. We interpret the period from t to $t + 1$ as one month, one quarter or one year as the case maybe.

We are interested in the growth rate of the aggregate consumption $C(t)$ over $[t, t + 1]$, namely, r(t) defined as $(C(t+1) - C(t))/C(t)$. Our goal is to derive the probability distribution of r.

The growth of aggregate (or macro) consumption arises from the aggregation of growths of the N individual (or micro) consumption. We assume that this micro growth occurs as a result of a (large) number of elementary events. The number of elementary events within a period (namely, over $[t, t + 1]$) is τ. One elementary event is that a consumer, consumer i, say, being randomly chosen from the set $\{1, 2..., N\}$ between t and $t + 1$, changes his/her consumption. We use the term "consumer," but it can be any micro unit such as a firm. We will later briefly discuss firm investment in the final Sect. 7.5.

A consumer may be chosen either uniformly with probability $1/N$ or with some other probabilities possibly dependent on the size of the initial level of consumption. If consumer i is chosen, $c_i(t)$ grows by a random factor λ:

$$c_i'(t) = \lambda c_i(t) \tag{7.40}$$

Here, $c_i'(t)$ stands for c_i immediately after elementary growth, not the time derivative of $c_i(t)$. At this event, no other consumer ($j \neq i$) experiences growth.

For simpler exposition, we assume that

$$\lambda = 1 + g, \quad \text{for} \quad \forall i,\ t \tag{7.41}$$

where

$$g = \pm\gamma \quad (0 < \gamma < 1). \tag{7.42}$$

Note that this is a particular type of multiplicative process for c_i. As we mention later, the probability distribution of λ does not matter at all.

It is important to keep in mind the difference between t and τ. One is the calendar time, t, and the other is τ, the number of elementary (micro) events within a given period of time. We denote the resultant growth rate of aggregate consumption as $r\ (t;\ \tau)$. That is, $r\ (t;\ \tau)$ is the growth rate of $C(t)$ between t and $t + 1$ when the number of elementary micro events during the period is τ. Although τ can be a random number, for simplicity, we use its expected value and denote it by τ.

Given τ, we can write the rate of growth of aggregate consumption as

$$r(t;\tau) = \sum_{i,k} g_{i,k} \qquad (k = 1, \ldots \tau) \tag{7.43}$$

Here,

$$g_{i,k} = \frac{(c_{i,k+1} - c_{i,k})}{C} = \frac{\pm\gamma c_i(t;k)}{C} \tag{7.44}$$

Here, $g_{i,k}$ indicates the k-th elementary growth that has occurred to $c_i(t)$. The total change that defines the growth rate of aggregate consumption $C(t)$ is the sum of these elementary growths that occurred to $c_1 \ldots c_N$. The total number of elementary events that have occurred is equal to τ.

In this model, the size of a jump of a micro unit is constant [Eq. (7.41)]. Thus, the micro behavior is described by the ordinary random walk. However, such micro growths occur τ times within a period. As a consequence, the growth of *aggregate* consumption $C(t)$ follows the *truncated Lévy flight* explained in Sect. 7.2 because τ is finite.

We make an important assumption that there is a lower bound constraint on the elementary micro growth process. That is, the level of consumption after elementary (micro) growth must be above the minimum, $c_{min}(t)$, defined by

$$c_{min}(t) = qc_{av}(t), \quad (0 < q < 1) \tag{7.45}$$

where

$$c_{av}(t) = \frac{C(t)}{N} \tag{7.46}$$

Here, q is the fraction of the average consumption that serves as the lower bound to all of c1 s. Thus, we actually obtain $c_i'(t)$ not as (7.40) but as

$$c_i'(t) = \max\{\lambda c_i(t), c_{min}(t)\} = \max\{(1 \pm \gamma)c_i(t), qc_{av}(t)\} \tag{7.47}$$

By scaling c_i by c_{av} (t), we define the fraction $y_i(t)$:

$$y_i(t) = \frac{c_i(t)}{c_{av}(t)} \tag{7.48}$$

By construction, it satisfies the normalization that the average of the fractions is 1. By changing variables into

$$Y_i = \log y_i \tag{7.49}$$

we observe that the basic dynamics become a kind of random walk with varying step sizes, i.e., a *truncated Lévy flight* with a lower reflecting barrier:

$$Y_i'(t) = Y_i(t) + \log \lambda \tag{7.50}$$

Here, again, the prime indicates the value after one elementary event, not the calendar time derivative.

We can easily derive the master equation. It is shown by Levy and Solomon (1996) that the asymptotic stationary distribution of Y, $P(Y)$ is exponential. Denote the exponent of this exponential distribution α. Then, we can derive R $(r; \tau)$ as the cumulative distribution function of r, that is, the probability that the growth rate of aggregate consumption $C(t)$ is less than or equal to r given τ elementary growths. For analytical details, the interested reader is referred to Aoki and Yoshikawa (2007, Ch.10).

There are many ways in which the growth rate of *aggregate* consumption r is realized.[9] The same aggregate growth rate, r, may be due either to a small number of elementary microgrowths with large step sizes, such as $r/2$ and $r/3$, or to a large number of small microgrowths, each with a small step size. This actually makes a difference to the emerging probability distribution of r. We can derive an exponential distribution for the aggregate growth rate r under the condition that the number of elementary events τ does not exceed a critical level, $\bar{\tau}$. This upper bound is defined by

$$\bar{\tau} = \left(\frac{N}{bq}\right)^{\alpha} \tag{7.51}$$

It depends on N, q, α and b. Here, b is defined by the following equation:

---

[9] A positive r and a negative r can be treated in almost identical ways. We focus on positive r.

$$r = bk \times \frac{\gamma}{b} \qquad (b = 1, 2, \ldots) \tag{7.52}$$

We assume that there are bk elementary events out of τ with magnitude γ/b each and that the rest of τ, namely, $\tau - bk$ events, makes almost no net contribution to the growth of the aggregate of consumption, r. A greater b means a greater number of elementary events each with a smaller contribution to make the growth rate r. We can then show that when $\tau < \overline{\tau}$ is satisfied, then the probability distribution of the growth rate r of the aggregate consumption $C(t)$ is exponential.

In summary, to obtain an exponential distribution for the growth rate of aggregate consumption, the number of elementary events within a given calendar time period τ cannot exceed a critical level $\overline{\tau}$ defined by Eq. (7.51). The probability density function of r has the following exponential distribution:

$$f(r, b) \propto \exp\left[-\frac{b}{\gamma} \log\left(\frac{\overline{\tau}}{\tau}\right) r\right] \tag{7.53}$$

Exponential distribution for the growth rate of the aggregate real variable is obtained when the number of micro growths within a short period of time τ is sufficiently small.

Financial Markets

We next study changes in asset prices using the same model. There are N investors or assets each with financial resources or wealth, $w_i(t)$:

$$W(t) = w_1(t) + \cdots + w_N(t) \tag{7.54}$$

As in the real-sector model, one of N stocks is randomly selected for an elementary event, i.e., a micro change. This random selection can be uniform with probability $1/N$ or can be modified to favor great asset values or wealthy investors. There are τ such micro or elementary events within a unit interval of time.

When asset i is selected, it undergoes the change

$$w_i'(t) = (1 + g) w_i(t) \tag{7.55}$$

where g is $\pm\,\gamma$ as in (7.42).

Again, $w'_i(t)$ indicates the value of w_i immediately after elementary growth in the i-th asset or agent's wealth. It is *not* the time derivative of $w_i\ (t)$.

The rate of return on financial assets or wealth over a calendar period from t to $t + 1$, $r(t)$, is defined, analogously as the rate of growth of the aggregate consumption, by

$$r(t) = \frac{W(t+1) - W(t)}{W(t)} \tag{7.56}$$

We are interested in the case where the probability distribution of r becomes a power distribution.

When there are τ elementary events during $[t, t+1]$, r is denoted by $r(t, \tau)$. By definition, $r(t, \tau)$ is as follows.

$$r(t, \tau) = \sum_{i,k} f_{i,k} \qquad (k = 1, \ldots, \tau) \tag{7.57}$$

where

$$f_{i,k} = \frac{\pm\gamma w_i(t; k)}{W(t)} \tag{7.58}$$

Again, the probability distribution of r depends on parameters τ and γ. Huang and Solomon (2001) demonstrate that depending on the parameters, a power law distribution emerges. There are a set of parameters for which power laws *with the exponent* α *close to* 3 emerge[10] in the financial model. The boundaries of the region are determined by two curves on the τ—r plane. The interested reader is referred to Aoki and Yoshikawa [2007, Ch.10]. When τ is larger than a critical value, a power low distribution of the growth rate of the value of financial assets is obtained.

In summary, in a truncated Lévy flight model in which the aggregate growth rate, r, is composed of a number of micro or elementary growths within a unit interval of time, the probability distribution of the growth rate of *aggregate* variable r depends crucially on the number of such micro events, τ. Specifically, when the frequency τ of micro changes in a given time is small, an exponential distribution is produced, whereas when the frequency exceeds a certain level, a power distribution is produced. Intuitively, this means that the frequency of change is high, and "violent" changes produce a power distribution.

Yamada et al. (2009) is another model. It is a simulation model to explain the power distribution of asset prices. The model considers two traders with a random walk for the reservation prices of the "sell" and "buy" prices. If the selling price is lower than the buying price, one unit of stock will be sold. The price is determined by the midpoint between the offer price and the bid price. If we take these assumptions literally, they are obviously too simple. However, this is not a flaw in the model. With just a few modifications to such a simple model, we can successfully explain the distribution of price changes and the distribution of intervals between transactions (Model 3 in the original paper).

[10] Huang and Solomon (2001) demonstrate by their simulations that the exponent α becomes close to 3.

7.5 Rationality in Financial Markets

When considering "bubbles", the biggest problem is the rationality of *ex ante* expectations. However, it is not wise to get too hung up on them when thinking about the problems of asset markets. Just remember the Fed View fallacy of US macroeconomists in the run-up to the 2008 financial crisis. It is pointless to pry into how "rational" the behavior of each individual pedestrian was in a disastrous human avalanche in which people fell down one after another. The fact remains that the disastrous accident occurred. The same is true for congestion on highways. In fact, physicists have studied the phenomenon of congestion on highways from such an obvious standpoint and have made considerable progress. We must reconsider the "rationality" of the financial market.

First, market participants are not truly interested in fundamental values. In Chap. 12 of Keynes's *General Theory*, this problem is discussed in its clearest form. Keynes's metaphorical "beauty contest" is so well known.

> Professional investment may be likened to those newspaper competitions in which the competitions have to pick out the six prettiest faces from a hundred photographs, the prize being awarded to the competitor whose choice most nearly corresponds to the average preferences of the competitions as a whole: so that each competitor has to pick, not those faces which he himself finds prettiest, but those which he thinks likeliest to catch the fancy of the other competitors, all of whom are looking at the problem from the same point of view. It is not a case of choosing those which, to the best of one's judgment, are really the prettiest, nor even those which average opinion genuinely thinks the prettiest. We have reached the third degree where we devote our intelligences to anticipating what average opinion expects the average opinion to be. And there are some, I believe, who practice the fourth, fifth, and higher degrees (Keynes 1936, 156).

Thaler, who won the 2017 Nobel Prize in Economics for his work in behavioral economics, believes that Keynes's beauty contest analogy remains an apt description of how financial markets work. In his book *Misbehaving* [Thaler 2015, Ch. 21], he explains an interesting "contest" he conducted in the *Financial Times* to recreate Keynes's beauty contest: "Guess a number 0 to 100 with the goal of making your guess as close as possible to two-thirds of the average guess of all those participating in the contest". Try it yourself, first, and it is a fun to see the distribution of FT reader guesses!

The Keynes's beauty contest is not a mere "mental gymnastics". Milton Friedman's monetarism was a major influence on US financial markets in the 1970s, and money stock (then known as money supply, M1) was the most closely watched economic variable. Every week, the Fed released a preliminary estimate of M1, but the weekly M1 was extremely inaccurate and was revised substantially afterwards. Only approximately half of the financial institutions were members of the Federal Reserve System at the time, and their reports to the Fed were made only on a voluntary basis. Without modern ICT, they were liable to lag. The magnitude of *ex post* revisions (standard deviation) was also available so that any movements within two standard deviations were not statistically significantly different from zero.

Nevertheless, the TB market was also very sensitive to small (i.e., within two standard deviations) movements. The increase in the money stock, which according to monetarism should fuel inflation, was interpreted in the market as a sign of near future tightening by the Fed concerned with inflation; market participants believed the market was telling a story that would lead to higher interest rates. Thus, the "rationality" of financial markets is just the kind of "rationality" that needs to be bracketed.

However, apart from this, there is a fundamental difference between the efficiency of asset markets and the efficiency of the real economy. The difference can be easily understood. In asset markets, one second matters, and market participants must stick to one second to act "rationally". In contrast, in the real economy, where consumers and firms consume and invest, one second does not matter. For consumers, having to worry about one second must be a psychological burden and create disutility.

There is another important difference. In the real economy, consumers and firms consume and invest for their own sake. In contrast, trading in financial markets is often for the purpose of "resale". This is also a factor making the difference of efficiency in financial markets and the real economy. In the real economy, efficiency is anchored to preferences and technology, whereas in the financial market, it is not.

It is reasonable for a firm to disregard transitory information that may be revised in months, let alone seconds, given the time required for capital investment and relevant horizon. In Tobin's *q theory*, a standard theory of corporate investment with adjustment costs, the ratio q of the stock price to the reproduction cost of the capital stock is a sufficient statistic to determine investment (Tobin 1969; Yoshikawa 1980; Hayashi 1982). However, in the empirical analyses, the q investment function derived from standard theory performs poorly. It is known that corporate investment is highly correlated with real variables such as sales and profits, not with q.

q is greatly affected by the stock price in the numerator, and stock price responds "rationally" to temporary information, in some cases, even to transitory information that is not related to fundamentals. Ueda and Yoshikawa (1986) show theoretically that firms must "rationally" ignore such temporary information when making investment decisions because unlike financial assets, real capital stocks, such as plants, machinery, technology and human resources, cannot be changed instantaneously. As a result, a firm's investment is not highly correlated with stock price or q but with real variables such as profits and sales. Insofar as stock price is "rationally" determined in the market and firms also "rationally" make investment decisions, this fact shows that efficiency in financial markets and efficiency with respect to real investment decisions are two different things.

One concrete policy proposal based on this understanding is the famous Tobin tax. It has been rather unpopular because it only disrupts short-term financial transactions. However, it is the short-term international capital movement that disturbs the economy. The Tobin tax is a simple fixed-rate tax on international capital flows with the desirable feature of heavy taxation on repetitive, high-frequency transactions and only negligible taxation on low-frequency transactions with a long-term horizon. Suppose that a single foreign currency transaction (a round-trip transaction of country A to country B and back to country A) is taxed at a rate of 0.2%. For a trader that makes this transaction every day, the effective annual tax rate would be 48%. However, the tax rate for every weekly transaction would be 10%, 2.4% for every monthly transaction, and the annual tax rate 0.2% for once a year transaction. Thus, short-term capital movements that take place at a high frequency are taxed heavily. It is estimated that 80% of foreign currency transactions are round trips of one week or less, and many of these transactions are done from a short-term perspective. Tobin offers the following episode.

> An undergraduate student of mine, upon graduating, got a job in the Chicago mercantile exchange and became assistant and apprentice to an active trader who had been an economics professor. After a few weeks, the young man made bold to ask his mentor about the long-run calculations that must – necessarily he thought – govern his trades. The reply was 'Sonny, my long-run is the next ten minutes.' (Tobin 2003, 42)

The Tobin tax is intended to discourage participants in Keynes's beauty contest. The basis of the proposal is the recognition that efficiency in the real economy is not the same thing as efficiency in financial markets.

In neoclassical theory, the allocation of real resources such as consumption and investment and asset prices are determined simultaneously. It does not take into account any difference between the real economy and financial markets. Therefore, the efficiency of asset markets is defined on the basis of consumer utility, and the same concept of Pareto optimality as in Walras' general equilibrium theory is applied. However, efficiency in the real economy and efficiency in financial markets are fundamentally different.

Summary

There is no doubt that a well-organized financial market makes a significant contribution to the healthy development of the real economy. However, it is a mistake to assume that all financial technologies, or even market movements, always have a positive impact on the real economy (Schularick and Taylor 2021). As we explained, the "efficiency" of financial markets and the efficiency of the real economy are fundamentally different. Ironically, the gap between the two was small when the technology involved in finance was still underdeveloped, but the gap has only grown as technology has advanced. Supercomputer-based financial transactions no longer have any relationship to the efficiency of the real economy. Whether rational or not, large asset price changes, particularly crashes, significantly disrupt the real economy. History has repeatedly demonstrated that Keynes was right in this view.

In finance, the profession appears to move away from the efficient markets theory. Even before the 2008–2009 global financial crisis, Shiller (2003) argued as follows:

> Academic finance has evolved a long way from the days when the efficient markets theory was widely considered to be proved beyond doubt. Behavioral finance—that is, finance from a broader social science perspective including psychology and sociology—is now one of the most vital research programs, and it stands in sharp contradiction to much of efficient markets theory.
>
> In further research, it is important to bear in mind the demonstrated weaknesses of efficient markets theory and maintain an eclectic approach. While theoretical models of

> efficient markets have their place as illustrations or characterizations of an ideal world, we cannot maintain them in their pure form as accurate descriptions of actual markets.
>
> Indeed, we have to distance ourselves from the presumption that financial markets always work well and that price changes always reflect genuine information. Evidence from behavioral finance helps us to understand, for example, that the recent worldwide stock market boom, and then crash after 2000, had its origins in human foibles and arbitrary feedback relations and must have generated a real and substantial misallocation of resources. The challenge for economists is to make this reality a better part of their models (Shiller 2003, 83, 102).

Andrew Lo's (2017) adaptive markets hypothesis is on this track, although the biological theory of evolution seems far-fetched. When Lo talks about physics, it is Newtonian classical physics. It is statistical physics that provides finance with powerful analytical tools. In the domain of standard macroeconomics, Gennaioli and Shleifer (2018), based on the experiences of the 2008 financial crisis, present a different approach to expectations than rational expectations in financial markets, which may lead us to a crisis.

An array of research mostly done by econophysicists amply demonstrates that in financial markets, imitation, herding and positive feedback can lead endogenously to instability. From the viewpoint of macroeconomics, the recognition that irrational bubbles occasionally occur is important because they disturb the macroeconomy. The macroeconomy together with financial markets is not in the neoclassical equilibrium.

Chapter 8
Summing Up

Over the last 50 years, macroeconomics, once synonymous with Keynesian economics, has drastically changed to neoclassical economics. According to Lucas (1987, 107), "the term 'macroeconomic' will simply disappear from use and the modifier 'micro' will become superfluous." The term "macroeconomic" survives in a hollow way, for the modern microfounded macroeconomics is nothing but a modified miniature of the Walrasian model of general equilibrium. It takes great pride in solid microfoundations, which Keynesian economics allegedly lacks.

However, the standard research program for microfoundations, in which optimization of microeconomic agents is explicitly analyzed, is on the wrong track. The reason is that in the macroeconomy, we never know the micro-constraint under which an economic agent optimizes. We cannot, therefore, seriously set up an optimization problems *for the purpose of macroeconomics*. All the microfounded models—Lucas rational expectations model, the labor search theory, and the more recent mean field game—rest on the fundamental assumption that economic agents differ if they differ only to the extent that realizations of relevant stochastic variables are different. They are assumed to share a common stochastic distribution as their constraints. This is wrong, not worthy of microfoundations. Alleged heterogeneity in these models is only superficial. They are simply stochastic versions of the representative agent models.

The structure of these models is vividly captured by the famous metaphor of islands due to Edmond Phelps. There, micro-agents are *symmetric*. However, this assumption is time and again rejected by empirical analyses of networks. For example, the analyses of corporate networks demonstrate that they are not random,but scale-free. As a result, we observe the Pareto distribution for firm size. This fact is fairly well known, but its significance for microfoundations for macroeconomics is not fully understood. This means that the assumption of standard microfoundations is wrong.

H. Yoshikawa, *Reconstruction of Macroeconomics: Methods of Statistical Physics, and Keynes' Principle of Effective Demand*, Advances in Japanese Business and Economics 3, https://doi.org/10.1007/978-981-19-5264-7_8

Keynes in his *General Theory* made the following remarks:

> Our criticism of the accepted classical theory of economics has consisted not so much in finding logical flaws in its analysis as in pointing out that its tacit assumptions are seldom or never satisfied, with the result that it cannot solve the economic problems of the actual world (Keynes, 1936, 378).

This applies to modern microfounded macroeconomics as well. Economic agents optimize more or less, but in the macroeconomy, we never know the micro-constraints under which they optimize. The situation is fundamentally different from "micro problems" such as industrial organization and the "market for lemons" in which economic agents may well share the same constraints. Thus, macroeconomics requires different methods than microeconomics,—methods of statistical physics.

It is a contribution of equilibrium search theory to demonstrate that allocation of labor is done through the matching of workers and firms under imperfect information in segmented markets. Unfortunately, search models are all stochastic representative agent models in which workers and firms are assumed to share a common probability distribution. Chapter 3, using methods of statistical physics, shows that matching functions routinely taken as given on the supply side, depends crucially on the level of real aggregate demand. Real aggregate demand affects the allocation of workers and thereby determines not only unemployment but also the distribution of labor productivity. This is the working of the labor market behind Keynes's principle of effective demand. The model of Chap. 3 provides Keynesian economics with the correct microfoundation, which does not depend on the details of the micro behaviors at all.

It was Keynes's insight into the working of the macroeconomy in which real aggregate demand is the fundamental constraint. *Aggregate demand becomes an effective constraint on the macroeconomy because it changes much more swiftly than technology or endowment*. Modern microfounded macroeconomics has purged real aggregate demand from theory. It can afford to do so because prices are assumed to clear all the markets in a Walrasian way. However, the economy in which we happen to live is *not* built that way.

Prices *do* play an important role in adjusting demand and supply, but *only locally, not all at once in the economy as a whole*, as Walras envisaged the macroeconomy. We, including economists, all observe every day that prices significantly affect the economic behavior of consumers and firms, but what we observe are the workings of prices in local markets. The Walrasian general equilibrium is a wholly different matter.

For example, the price of high-quality steel is typically determined by bilateral negotiations between large steel and automobile companies. Information relevant to both parties is only a small subset of the "price vector" in the Walrasian general equilibrium model and includes *other than prices*. Producers of consumables are keen on market share and strategically determine their prices in their respective markets. In this way, prices are determined in a chain of local markets. *Fairness* plays a key role in the determination of prices. Our analysis in Chap. 6 indeed demonstrates that prices change in clusters from upstream to downstream sectors. These changes we observe in the actual world are far from the Walrasian fashion.

> In the absence of instantaneous and complete market clearing, output and employment are frequently constrained by aggregate demand. In these excess-supply regimes, agents' demands are limited by their inability to sell as much as they would like at prevailing prices. Any failure of price adjustments to keep markets cleared opens the door for quantities to determine quantities, for example, real national income to determine consumption demand, as described in Keynes' multiplier calculus (Tobin, 1993, 46).

The role of real aggregate demand is not confined to the short run. A succession of short runs leads us to the medium run, and the long run is nothing but a succession of the medium runs. Schumpeter even rejected the notion of the long-run trend. In contrast to modern economists who tend to take the short-run fluctuations as a perturbation to steady growth, he wrote in his magnum opus *Business Cycles* that "real is only the cycle itself" (Schumpeter, 1939, 207).

There is nowadays a firm consensus that long-run economic growth is determined only by supply-side factors. However, growth is also strongly conditioned on demand because in advanced countries, *saturation of demand* is the fundamental factor constraining economic growth, and demand-creating innovations are the engine of growth.

Economists endeavor to endogenize innovations and the accumulation of human capital as the outcome of economic agents' optimization in growth models. However, the eminent economic historian Postan reviewing the history of technical progress concludes as follows:

> In this way, an historical account of technological change, multivariable as all such accounts are bound to be, resolves itself into an entire gamut of economic and social factors, those of supply and those of demand (Postan, 1971, 143).

Technical progress and innovations depend plainly on so many factors, many noneconomic, that such efforts seem not so promising. One thing important is that they depend crucially on demand.

Meanwhile, financial markets play a central role in allocating investment, which directs the future of our economy. They are very competitive and in normal times efficient. However, occasionally more or less irrational herding and positive feedback lead markets to instability. In such abnormal regimes, the anchor of the fundamental price becomes shaky, and bubbles or crashes occur. Bubbles damage the macroeconomy. Thus, the efficient market hypothesis must be discarded as universal theory. It is dangerous to the extent that it distracts us from our careful attention to possibly irrational markets. It is the great achievement of econophysics to clarify the broad picture of financial markets.

Although methods of statistical physics clarify the workings of financial markets and provide the principle of effective demand with microfoundations, we do not have any self-contained mathematical model ready for explaining the macroeconomy and prescribing policies because the macroeconomy is an open system in the sense that aggregate demand changes for so many different reasons, and "narrative" bearing bubbles are also different in time and place (Skidelsky (2020)). Keynes left the following remarks on the nature of economics in his obituary on his master Alfred Marshall.

> The study of economics does not seem to require any specialised gifts of an unusually high order. Is it not, intellectually regarded, a very easy subject compared with the higher branches of philosophy and pure science? Yet good, or even competent, economists are the rarest of birds. An easy subject, at which very few excel! The paradox finds its explanation, perhaps, in that the master-economist must possess a rare *combination* of gifts. He must reach a high standard in several different directions and must combine talents not often found together. He must be mathematician, historian, statesman, philosopher – in some degree. He must understand symbols and speak in words. He must contemplate the particular in terms of the general, and touch abstract and concrete in the same flight of thought. He must study the present in the light of the past for the purposes of the future. No part of man's nature or his institutions must lie entirely outside his regard. He must be purposeful and disinterested in a simultaneous mood; as aloof and incorruptible as an artist, yet sometimes as near the earth as a politician. Much, but not all, of this ideal many-sidedness Marshall possessed. But chiefly his mixed training and divided nature furnished him with the most essential and fundamental of the economist's necessary gifts – he was conspicuously historian and mathematician, a dealer in the particular and the general, the temporal and the eternal, at the same time (Keynes, 1924, 173—174).

Keynes's remarks apply more to macroeconomics than to microeconomics. A master of game theory might offer a practical solution to a *microeconomic problem* using a mathematical model, but the same is not expected for *macroeconomic problems*. Microeconomic problems rarely become historical events, but macroeconomic troubles such as the Great Depression during the 1930s the global financial crisis and its aftermath, and COVID-19 in this century all have left undeniable marks in history. As such, they require more than simple ready-made solutions based on mathematical models. That is what Keynes meant.

This does not mean that macroeconomics is useless. We contend that the methods of statistical physics must replace the standard microfoundations based on optimization exercises. In any case, the role of real aggregate demand is essential. Aggregate demand not only determines short-run fluctuations but also affects growth. The troubles with financial markets also affect the macroeconomy mainly through real aggregate demand. The causes of fluctuations of aggregate demand are so diverse, sometimes even unique in history as the former Federal Reserve Board Chairman Alan Greenspan called the financial crisis during the years 2008–2009 "once-in-a-century," that the problems cannot be solved by ready-made policies prescribed by simple mathematical models. That is precisely why the principle of effective demand is so important. However, in the name of microfoundations, the profession has purged aggregate demand from macroeconomics and wandered without being able to understand and solve the economic problems of the actual world. Macroeconomics must change.

Appendix
Pioneers

The statistical physics approach to macroeconomics advanced in this book has yet to gain firm ground in economics. The fact is that a majority of economists do not pay any attention to it. But if we go back in history, we can find the footprints of the pioneers inside and outside of economics. This appendix briefly reviews the history albeit it is necessarily partial.

The biggest obstacle to the application of methods of statistical physics to economics is that many economists believe that there is an essential difference between the motion of inorganic particles and purposeful human economic activity. In this regard, we must understand that the motion of inorganic particles and molecules is also a solution to the variational problem (variational principle), as explained in Chap. 2. It is not surprising that those who discovered the variational principle in the eighteenth century found in it a "God" as the "subject" who optimizes the motion of inorganic objects. In light of this fact, for the third-party observer, there is no essential difference between the movement of inorganic particles/molecules and a Ramsey-like consumer. What is essential is the number of microscopic subjects. In other words, when the number of micro agents is large, it is meaningless to study the movement of microscopic subjects in detail, and methods of statistical physics must be used. Here lies the essential difference between macroeconomic phenomena and microeconomic problems. For the latter, the motives and optimization of economic agents are important.

One of the pioneers who articulated these ideas almost ninety years ago was the physicist Torahiko Terada, who wrote the following in response to the mainstream academic community of his time. Terada had in mind biology, not economics, but he wrote against the mainstream of the academic world at the time which held that there is no way that physics can contribute to our understanding of living things.

> When you make a statistical study of such a high class of animals as human beings, it often makes no difference if you ignore the freedom of will of each individual human being and consider them as a simple group of inorganic matter. The statistics on the "density" and "average speed" of human beings walking on the streets, for example, apply the same formula as to the density of colloidal particles. It has been confirmed that it is appropriate to use it. …

H. Yoshikawa, *Reconstruction of Macroeconomics: Methods of Statistical Physics, and Keynes' Principle of Effective Demand*, Advances in Japanese Business and Economics 3, https://doi.org/10.1007/978-981-19-5264-7

> It is truly regrettable that such research is often regarded as heretical because of the canonical belief that biology cannot be understood by physics. What impedes the progress of science is not the lack of understanding of laymen, but rather the lack of awareness by scientists of the mission and nature of science itself. (Torahiko Terada, "Fauna seen as substance group," in *Science*, 1933, later in *Matter and Language*, Tokyo: Iwanami Shoten, 1935, 134, 138)

As for the application of physics to biology defended by Torahiko Terada, biophysics now has been well established. Nevertheless, even if one succeeds in biology, organisms and humans are different. Humans with well-developed brains are capable of making highly rational decisions. The majority of economists believe that.

However, there are many examples in which methods of statistical physics have been effectively applied to human behavior. A good example is the analysis of vehicular congestion on highways. The subject is not inorganic particles, but humans. A driver certainly optimizes albeit in a trivial way in order to get to his or her destination as soon as possible under the constraint of safety. If there is only one car on the highway, it is enough to analyze the micro behavior of the driver. However, when the number of vehicles increases, the qualitatively different macro phenomenon, i.e., traffic congestion, occurs on the highway. Here, methods of statistical physics are most effective (Montroll, 1987; and Bando, 2009). Thus, the effectiveness of statistical physics is not determined by whether the micro subjects are inorganic particles or human. The method is effective for analyzing the collective phenomena where the number of microscopic subjects is large.

Champernowne's Study of Income Distribution

In fact, statistical approaches have been very successful in explaining the empirical law or regularity found in economic phenomena. A famous successful example is the study of income distribution by Champernowne (1953).

Pareto (1896) found that in many countries the distribution of income, especially for the upper-income group, follows power laws. It is now known that the power distribution is found universally in natural phenomena, but one found by Pareto is believed to be the oldest in the history of the discovery of empirical power distribution. For this reason, Pareto is perhaps the most widely known economist today among physicists who are interested in econophysics and power distribution; Pareto is famous only because of the "Pareto distribution", and no physicist knows the concept of the Pareto optimum!

Champernowne's work began when he was instructed by his tutor Keynes while he was a student at Cambridge University. Keynes's suggestion was to make a mathematical model that would explain how the income distribution follows the Pareto distribution. This research was subsequently submitted as a thesis for a fellowship at King's College in the autumn of 1936. Part of the study was published in the *Economic Journal* after the Second World War (Champernowne, 1953), but the entire manuscript was published in book form in 1973 after a lapse of more than 30 years (Champernowne, 1973).

This study analyzed the empirically stable pattern found in the income distribution (the Pareto distribution) using a Markov model. The multiplicative stochastic process explained in Chap. 7 plays an essential role in the theoretical model that explains the Pareto distribution.

Income Y varies based on the following multiplicative stochastic process.

$$Y_{t+1} = r_{t+1} Y_t \tag{A.1}$$

r *is a* growth factor (growth rate plus 1). If r is greater than 1, income Y rises, and conversely, if r is less than 1, Y falls. r is a random variable with identical and independent distribution. The density function $f(r)$ ($r \geq 0$) does not depend on the level of Y. In other words, the "growth rate" follows the same probability distribution regardless of the level of income. This assumption plays an essential role in obtaining the power distribution.

In this model, we consider the distribution of income Y, specifically, the probability that income Y is greater than a certain level x

$$G_t(x) = P(Y_t \geq x) \tag{A.2}$$

The evolution equation for the distribution G_t is as follows:

$$\begin{aligned} G_{t+1}(x) = P(Y_{t+1} \geq x) &= P\left(Y_t \geq \frac{x}{r} \text{ and } r_{t+1} = r\right) \\ &= \int_0^\infty G_t(\frac{x}{r}) f(r) dr \end{aligned} \tag{A.3}$$

Therefore, if there is a time-independent stationary distribution G, it satisfies

$$G(x) = \int_0^\infty G\left(\frac{x}{r}\right) f(r) dr \tag{A.4}$$

The equation where $G(x)$ is the power distribution is

$$G(x) = \frac{A}{x^\alpha} \quad (A > 0) \tag{A.5}$$

Substituting (A.5) into (A.4), we obtain

$$\int_0^\infty r^\alpha f(r) dr = E(r^\alpha) = 1 \tag{A.6}$$

Thus, the power distribution is a stationary distribution of multiplicative stochastic processes, and its order α is determined in such a way that α-th moment of the growth factor r is 1. Gabaix (2009) calls the equation to determine the order α of power distribution, (A.6), the "Champernowne equation". This equation is nothing but Eq. (7.23) of the Kesten/Goldie theorem on the multiplicative process explained in Chap. 7.

Why did Champernowne come up with this stochastic model in the first place? According to the standard neoclassical theory, individual income is determined by the amount of factors of production such as land, financial assets, and the ability to generate labor income the individual owns, and prices of these factors of production determined in markets. Champernowne calls the various factors of production owned by each person, particularly labor, "qualifications". He did not use the usual term "labor", but rather "ability", as a qualification to determine an individual's labor income. Champernowne (1973, 14), however, studied at Cambridge University, the mecca of neoclassical economics as exemplified by Marshall and Pigou. The theory of "marginal productivity" for the determination of factor prices was the ABC of economics. What is the relationship between his theory, which he calls "a stochastic-equilibrium model", and the neoclassical theory?

This question was apparently asked to Champernowne in his oral examination for a fellowship at King's College. His answer was as follows. Although forces of marginal productivity explain how the price of each factor of production is determined in the marketplace, individuals actually own a number of factors of production (Champernowne's so-called qualifications). The simple term "labor" does not mean that there is one type of labor that can be reduced to a quantity. The quality of labor varies widely, and the labor supplied by each person is fundamentally different (hence the qualifications). This is the reason that the matching in the labor market explained in Chap. 3 is so complex that it is impossible for a third party to know the real micro situations of both workers and firms.

Besides, both the endowment and prices of the production factors owned by individual vary in different ways. These changes are again so complex that attempts to identify their causes theoretically and then to explain the changes in individual income by adding them together are bound to fail. This is similar to the complexity of the micro-environment surrounding the search and matching of workers and firms in the labor market described in Chap. 3. Given the complexity of the micro-environment that determines an individual's income, it is much more productive to explain the macro income distribution by assuming that an individual's income is determined by the stochastic changes that have occurred during the period given a person's initial income, namely, applying a Markov model. This was Champernowne's idea.

The income distribution (Pareto distribution) of a country is a macro phenomenon. Behind the macroscopically stable income distribution, individual incomes are constantly changing. The factors that determine such fluctuations are, as we have repeatedly remarked, too complex and can only be described as random in the eyes of a third-party analyst. The only valid model is a stochastic model. The income distribution as a macroscopic phenomenon must be explained as stochastic equilibrium. This is what Champernowne thought.

Champernowne and Gibrat (1931) are representative of the stochastic approach in the prewar period.[1] After the war, the Markov model came to be used to explain a variety of economic and social variables such as urban population and frequency of alphabetic use. Particularly famous is the work of Zipf (1949), who found that various economic–social variables, such as the population of a city, the frequency of alphabet use, and so on follow power distribution of order 1:

$$P(s) = P_r\{S \geq s\} \sim \frac{1}{s} \tag{A.7}$$

$P(s)$ is probability distribution in which a random variable S (usually size or frequency) is greater than s. The power distribution of order 1 is called the "Zipf law".

The work of Ijiri and Simon (1979), conducted 30 years after Zipf, which attempted to explain the observed distribution of firm sizes (power distribution) by stochastic process, is widely known. As Champernowne's study made clear, it is a multiplicative stochastic process that plays an essential role in explaining empirical power distribution. Ijiri and Simon have also studied such processes, e.g., Ijiri and Simon (1975). Recent work on the underlying stochastic process that gives rise to power distribution is Saichev et al. (2010).

These studies such as Zipf, Champernowne, and Ijiri and Simon have been well known, but have failed to make deserved impacts on the mainstream of the profession. Sutton (1997), who surveyed these studies, argued that this was because they were "merely statistical and probabilistic" and did not have solid "micro-foundation". Micro-foundations refers, of course, to the optimization of microeconomic agents. However, the very starting point of the stochastic approach advanced by Champernowne and others is the recognition that for analyzing macro phenomena, standard micro-optimization exercises are futile.

The stochastic model that explains macroeconomic phenomena such as income distribution and firm size distribution is the correct "micro-foundation". The macro economy consists of a large number of micro-economic entities, and the economic environment surrounding each micro entity is a complex one that can only be perceived as random by a third party analyzing it. Champernowne's method was based on honest acknowledgment of this fact.

The "Russian School" of the 1930s

Another group that developed a statistical and probabilistic approach to economics before the Second World War was the "Russian School" in the 1930s. The Russian School was represented by Slutsky, who was famous for his "Slutsky equation", and Kondratiev, who was famous for his theory of the very long 50-year business cycle. The background of the Russian School must be the fact that Russian mathematics

[1] Gibrat's (1931) work predates Champernowne's work of 1936 by several years. Champernowne (1973, footnote 5) honestly confesses that he was somewhat disappointed when he learned of its existence after completing his own research without knowing of Gibrat's book.

had led the development of probability theory since the nineteenth century, as we know the names of Chebyshev, Markov, Kolmogorov, Khinchin, and others. Indeed, Slutsky was also originally a mathematical statistician.[2]

Slutsky has a famous study of business cycles. That paper, originally published in 1927 in the *Bulletin* of the Institute of Economic Research in Moscow and entitled "The Sum of Random Causes as a Source of Cycles", was translated into English and published in 1937 in *Econometrica*.[3] As the title suggests, the paper shows that the sum of random disturbances (moving average) is akin to a sine curve. Slutsky's study, along with the earlier work of Yule (1926), has been often taken as a warning that time series processed by moving averages produce "spurious cycles" (e.g., Frisch 1931). However, as Okada (2006, Chap. 6) points out, such an interpretation is incorrect. If one reads Slutsky's paper, it is clear that he did not merely write the paper as a warning against spurious cycles fabricated by moving averages, but he positively advanced one theory of the business cycle.

> Thus we are led to our basic problem: is it possible that a definite structure of a connection between random fluctuations could form them into a system of more or less regular waves? Many laws of physics and biology are based on chance, among them such laws as the second law of thermodynamics and Mendel's laws. But heretofore we have known how regularities could be derived from a chaos of disconnected elements because of the very disconnectedness. In our case we wish to consider the rise of regularity from series of chaotically random elements because of certain connections imposed upon them.
>
> Suppose we are inclined to believe in the reality of the strict periodicity of a business cycle, such, for example, as the eight-year period postulated by Moore. Then we should encounter another difficulty. Wherein lies the source of the regularity? What is the mechanism of causality which, decade after decade, reproduces the same sinusoidal wave which rises and falls on the surface of the social ocean with the regularity of day and night. It is natural that even now, as centuries ago, the eyes of the investigators are raised to the celestial luminaries searching in them for an explanation of human affairs. One can dauntlessly admit one's right to make bold hypotheses, but still should not one try to find out other ways? What means of explanation, however, would be left to us if we decided to give up the hypothesis of the superposition of regular waves complicated only by purely random components? The presence of waves of definite orders, the long waves embracing decades, shorter cycles from approximately five to ten years in length, and finally the very short waves, will always remain a fact begging for explanation. The approximate regularity of the periods is sometimes so distinctly apparent that it, also, cannot be passed by without notice. Thus, in short, the undulatory character of the processes and the approximate regularity of the waves are the two facts for which we shall try to find a possible source in random causes combining themselves in their common effect (Slutsky 1937, 20–21).

Slutsky did not merely warn against spurious cycles, but he advanced rather a new business cycle theory that suggests that the macro phenomenon which we call business cycles is generated by the accumulation of many random causes (moving averages). He presented a stochastic business cycle model (a simulation in today's

---

[2] See Okada (2006) on Slutsky's background and his relationship with Kondratiev.

[3] Slutsky's article in *Econometrica* was included by David Cass and Lionel W. McKenzie in *Selected Readings in Macroeconomics and Capital Theory from Econometrica* (MIT Press 1974). Around 1974, economists who understood the significance of the paper, rather than saying that it had no micro-foundations, were still in the center of the profession.

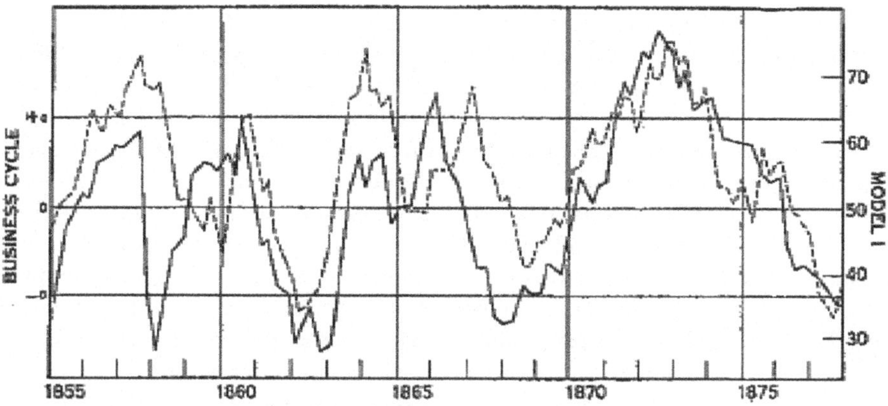

Fig. A.1 UK Business cycle index (1855–1875, *solid line, left axis*) and Slutsky's stochastic model I (*dashed line, right axis*). *Source* Slutsky (1937), Fig. 3

terminology) by using lottery numbers issued by the Soviet government to generate random numbers, and taking their moving averages. Figure A.1 compares the results of his simulation with the British business cycle index (1855–1875). Through such demonstrations, Slutsky asserted the validity of his probabilistic approach.

How did Slutsky come up with such a theory? It must have been the same idea that Champernowne had in explaining the Pareto law of income distribution. That is, there are innumerable "causes" of change in the economy, but there is no knowing what they are, and therefore, we have no choice but to regard them as random for the purpose of studying macro phenomenon. Of course, as Kuznetz (1929) pointed out, if Slutsky's stochastic approach is correct, then traditional business cycle theories mentioned in Chap. 4, which theoretically identify the "causes" of cyclical fluctuations, would have no place in economics. Do Samuelson's (1939), Hicks's (1950), and others' multiplier–acceleration models become meaningless? It is certainly too extreme to consider everything as random. However, it is notable that Slutsky proposed a new business cycle theory that considers stochastic disturbances as the root cause.

Kondratiev, a colleague of Slutsky's at the Moscow Institute of Economic Research, is famous for his work on the "long wave" of the 50-year cycle.[4] The "Kondratiev cycle" was immortalized in the history of economics by Schumpeter (1939) in his *Theory of the Business Cycle*. The Kondratiev cycle looks for the causes of economic fluctuations in wars and such major technological advances as the railroad. Thus, it is not a "stochastic approach" like Slutsky's. But on the other hand, there was actually a "probabilistic–statistical" methodology behind Kondratiev's empirical work.[5]

[4] Kondratiev was purged by Stalin at the young age of 46. On the other hand, Slutsky, who was born in 1880, lived until 1948. The relationship between Slutsky and the Institute for Economic Research, including the circumstances that separated their fates, is described in detail in Okada (2006, 178–182).

[5] The author learned this point from Okada (2006, 247–264); see also Klein (1999).

This methodology was most clearly described in his 1926 paper entitled "The Problem of Prediction" (Kondratiev, 1926). This lengthy paper is a little redundant at times, but it advanced an idea that the stable "laws" found in a macro system is based on the law of large numbers of probability theory (Sect. 4 of his paper). He repeatedly states that such methodologies have been successful in the natural sciences, and it is worth noting that he mentions Boltzmann's molecular kinetic theory of gases in his article. Kondratiev (1926) is a very early example of an economist referring to Boltzmann's statistical mechanics. In the natural sciences such as physics and chemistry, the number of microscopic components is so large (Avogadro's number 6×10^{23}) that the law of large numbers is strictly established, and therefore the laws found in the macroscopic system produce stable relationships worthy of the name of the law. In contrast, in the economy, the number of microscopic entities is relatively small so that the law can only be established imperfectly and fluctuations cannot be ignored. Nevertheless, Kondratiev argues that "despite the randomness of individual cases of both suicide and crime, surprisingly stable patterns can be found when viewed as social group phenomena. Such macro patterns must be captured by the "probabilistic–statistical method." He advanced "the view that a large number of random events are the basis of the regularities found in natural–social phenomena".

Micro-foundations of the Cobb–Douglas Function

Another example of the application of the stochastic approach to economics is the analysis of the aggregate Cobb–Douglas production function by Houthakker (1955–1956). Consider an economy or industry consisting of a large number of firms or production units. The production of firm i is characterized by a fixed input coefficient. For simplicity, let us assume that firm i uses two production factors 1 and 2 as inputs, and produces one output. The amount of inputs 1 and 2 required to produce one unit of output for firm i is a_1^i, a_2^i. The input factor a^i varies across firms. When the prices of output and the two inputs are p_0, p_1, p_2, respectively, we have

$$p_0 - p_1 a_1^i - p_2 a_2^i \geq 0 \tag{A.8}$$

Equation (A.8) guarantees that only those firms whose profits are not negative will produce. If the above inequality (strictly) holds, firms have incentives to increase production as much as possible, but fixed factors of production (e.g., management capacity) are the constraint, and all firms are given the upper limit of output capacity. Assuming that the capacity of firms with input coefficients a_1 and a_2 are continuously distributed, the density function can be expressed as $\phi(a_1, a_2)$, where capacity is expressed as

$$\varphi(a_1, a_2) da_1 da_2 \tag{A.9}$$

Taking into account this distribution of capacity and the constraint that profit is positive, the total output X_0 is given by

$$X_0 = \int_0^{p_0/p_1} \int_0^{\frac{p_0+a_1p_1}{p_2}} \varphi(a_1, a_2)da_1da_2 \tag{A.10}$$

In the same way, the total amount of the two inputs is

$$X_i = \int_0^{p_0/p_1} \int_0^{p_0+a_1p_1} a_i\varphi(a_1, a_2)da_1da_2 \quad (i = 1, 2) \tag{A.11}$$

Houthakker (1955–1956) shows that when the distribution of input coefficients a_1 and a_2 is a power distribution,

$$\varphi(a_1, a_2) = Aa_1^{\alpha_1-1}a_2^{\alpha_2-1} \quad (\alpha_i \geq 1) \tag{A.12}$$

then, the macro production function is

$$X_0 = CX_1^{\frac{\alpha_1}{\alpha_1+\alpha_2+1}} X_2^{\frac{\alpha_2}{\alpha_1+\alpha_2+1}} \quad (C \text{ is a positive.}) \tag{A.13}$$

That is, the macro production of function is of the Cobb–Douglas type.

In Houthakker's analysis, there is also variation in profits. It assumes that there is an upper limit to production capacity of each firm. On the other hand, Keynes's "Principle of Effective Demand" explained in Chap. 3 assumes that the demand for firms' output is quantitatively constrained. Although there is a difference between the two, the economy is not in neoclassical equilibrium in either model.[6]

Houthakker's analysis, which assumes heterogeneous firms with different efficiencies, marked an interesting first step, but its impact on macroeconomics was limited. Kazuo Sato (1974), following the lead of Houthakker's work, conducted research on the problem of aggregation of macro production function in the early 1970s. The time was, however, when macroeconomics had just entered an era of a "rational expectations revolution".

Tobin's "Stochastic Macro-equilibrium"

The Phillips curve—a negative correlation between the unemployment rate and inflation—which originated in the empirical work of Phillips (1958), was widely accepted by the profession in the 1960s, but Friedman (1968) argued that in the long run, the unemployment rate equals the "natural unemployment rate", which is determined independently of inflation. The "micro-foundation" of the unemployment rate and inflation by Phelps et al. (1970) gave rise to a major field of research, the "search theory" of the labor market. However, as discussed in detail in Chap. 3, the search theory was based on the neoclassical theory of equilibrium based on wrong micro-foundations.

---

[6] We can actually read φ, which in Houthakker's model is a density function of the capacity of the firm with input coefficients a_1 a_2, as the distribution of the effective demand of such a firm.

Meanwhile, Tobin (1972) proposed the concept of *stochastic macro-equilibrium* to show that there is a trade-off between the unemployment rate and inflation even in the long run—the long-run Phillips curve is downward-sloping rather than vertical. His concept of stochastic macro-equilibrium is as follows:

> It is stochastic in the sense that random inter-sectoral shocks keep individual labor markets in different imbalances. At the same time, it is macro-equilibrium in the sense that uninterrupted inter-market fluctuations produce fairly distinct aggregate movements (Tobin, 1972, 9).

This is exactly the same idea as that of Champernowne, who explained the Pareto distribution with the Markov Model. In fact, Champernowne also used the term "stochastic equilibrium".

Iwai's (1981) "disequilibrium dynamics" was an attempt to formalize Tobin's (1972) concept explicitly in a mathematical model. Iwai (1981) explains what Tobin (1972) called stochastic macro-equilibrium by using the metaphor of a swarm of mosquitoes.

> A macroscopic steady state is an *equilibrium of microscopic disequilibria*. Its characteristic feature lies in the marked contract between the tranquil and homogeneous aggregative behavior and the volatile and heterogeneous individual behaviors. To use a biological analogy (which is certainly a bad habit of economists), it would be likened to a swarm of mosquitoes which, observed from a distance, appears merely as a white cloud more or less stable in its figuration. But on closer examination (which we do taking some care not to be stung), it would be found to be a cluster of thousands of mosquitoes, each dancing wildly to and fro and up and down. Similarly, if we are content to examine only the aggregate behavior of the labor market, we will never understand the actual dynamic behavior of individual firms in it. If, on the other hand, we focus our attention only on the short-run behavior of an individual firm, we easily lose the sight of the structure of the labor market as a whole. Only if we succeeded in synthesizing the micro- and macroscopic analyses would a complete picture of the dynamic and complex labor market emerge before our eyes (Iwai, 1981, 210).

"Stochastic macro-equilibrium" is a statistical physics concept; the works of Tobin (1972) and Iwai (1981) were pioneering attempts in such a direction, but unfortunately, it never entered the mainstream of the profession.

Contributions of Physicists

Attempts to apply the methods of statistical physics to economic phenomena have been made mostly by physicists. The physicist Haken (1977) was an early pioneer of analyzing pattern formations in nature and society under the name of "synergetics". Chapter 11 of Haken (1977) is the model entitled "sociology" and analyzes public opinion formation. It can be easily applied to economic phenomena, and is, in fact, a precursor to the business cycle model of Aoki (1998) explained in Chap. 2. Montroll (1981, 1987), a renowned statistical physicist, also showed how entropy plays an essential role in the understanding of economic phenomena through a number of empirical analyses. In Japan, Fukao (1987) wrote a textbook on this subject. Weidlich (2000), a German physicist, also has made many contributions. Following the spirit, Schweitzer (2003) and Di Guilmini, et.al. (2017) have advanced agent based models.

In the 1980s, energetic empirical analyses were carried out by the physicist Stanley and his associates, and the name "econophysics" was finally established in 2000 with a focus on finance (Mantegna & Stanley, 2000; Takayasu 2005). Sornette (2003, 2014), Bouchaud and Potters (2003), and Roehner (2009) also have made substantial contributions to the analyses of financial markets with emphasis on bubbles and risk. Recent contributions by Japanese physicists include Aoyama et al. (2010a, 2017). All these efforts have been made by physicists primarily interested in socio-economic phenomena, and economists have been generally unresponsive. A notable exception is Foley (1994), who introduced the entropy maximization principle into general equilibrium theory, and proposed "statistical equilibrium theory".[7]

Econophysics broadly defined still retains its heretical status in economics. Under these circumstances, as a pioneer, Masanao Aoki actively advanced the probability–statistical approach in macroeconomics (Aoki, 1996, 2002; Aoki & Yoshikawa, 2007).

[7] See also Smith and Foley (2008). Foley (1994) analyzes the pure exchange economy. Toda (2009) introduces production to this model. These efforts are interesting but basically concern the Walrasian model of equilibrium.

References

Abowd, J., Kramarz, F., & Margolis, D. (1999). High wage workers and high wage firms. *Econometrica, 67*(2), 251–333.

Abowd, J., Haltiwanger, J., Jarmin, R., Lane, J., Lengermann, P., McCue, K., McKnney, K., et al. (2005). The relation among human capital, productivity, and market value: Building up from micro evidence. In C. Corrado, J. Haltiwanger, & D. Sichel (Eds.), *Measuring capital in the new economy* (pp. 153–204). University of Chicago.

Achdou, Y., Buera, F. J., Lasry, J.-M., Lions, P-L., & Moll, B. (2014). Partial differential equation models in macroeconomics. *Philosophical Transactions of the Royal Society A, 372*, 20130397.

Aghion, P., Antonin, C., & Bunel, S. (2021). *The power of creative destruction.* Harvard University Press.

Aghion, P., & Howitt, P. (1992). A model of growth through creative destruction. *Econometrica, 60*, 323–351.

Akerlof, G. (1970). The market for "Lemons": Quality uncertainly and the market mechanism. *Quarterly Journal of Economics, 84*(3), 488–500.

Akerlof, G., Dickens, W., & Perry, G. (1996). The macroeconomics of low inflation. *Brookings Papers on Economic Activity, 1*, 1–76.

Anderson, P. W. (1972). More is different. *Science, 177*(4047), 393–396.

Aoki, M. (1996). *New approaches to macroeconomic modeling: Evolutionary stochastic dynamics, multiple equilibria, and externalities as field effects.* Cambridge University Press.

Aoki, M. (1998). A simple model of asymmetrical business cycles: Interactive dynamics of large number of agents with discrete choices. *Macroeconomics Dynamics, 2*(4).

Aoki, M. (2002). *Modeling aggregate behavior and fluctuations in economics.* Cambridge University Press.

Aoki, M., & Shirai, Y. (2000). A new look at the diamond search model: Stochastic cycles and equilibrium selection in search equilibrium. *Macroeconomic Dynamics, 4*, 487–505.

Aoki, M., & Yoshikawa, H. (2002). Demand saturation/creation and economic growth. *Journal of Economic Behavior & Organization, 48*(2), 127–154.

Aoki, M., & Yoshikawa, H. (2006). Stock prices and the real economy: Power law versus exponential distributions. *Journal of Economic Interaction and Coordination, 1*, 45–73.

Aoki, M., & Yoshikawa, H. (2007). *Reconstructing macroeconomics: A perspective from statistical physics and combinatorial stochastic processes.* Cambridge University Press.

Aoki, M., & Yoshikawa, H. (2008). The nature of equilibrium in macroeconomics: A critique of equilibrium search theory. *Economics E-journal* (www.economics-ejournal.org) *Discussion Paper*, No. 2008-37.

H. Yoshikawa, *Reconstruction of Macroeconomics: Methods of Statistical Physics, and Keynes' Principle of Effective Demand*, Advances in Japanese Business and Economics 3, https://doi.org/10.1007/978-981-19-5264-7

Aoki, M., & Yoshikawa, H. (2012). Non-self-averaging in macroeconomic models: A criticism of modern micro-founded macroeconomics. *Journal of Economic Interaction and Coordination, 7–1*, 1–22.

Aoyama, H., Aruka, Y., & Yoshikawa, H. (2020). *Complexity, heterogeneity, and the methods of statistical physics in economics*. Springer.

Aoyama, H., Yoshikawa, H., Iyetomi, H., Fujiwara, Y. (2009). Labor productivity superstatistics. *Progress of Theoretical Physics Supplement*, No. 179, 80–92.

Aoyama, H., Fujiwara, Y., Ikeda, Y., Iyetomi, H., & Soma, W. (2010a). *Econophysics and Companies*. Cambridge University Press.

Aoyama, H., Yoshikawa, H., Iyetomi, H., & Fujiwara, Y. (2010b). Productivity dispersion: Facts, theory, and implications. *Journal of Economic Interaction and Coordination, 5*(1), 27–54.

Aoyama, H., Guilmi, C., Fujiwara, Y., & Yoshikawa, H. (2022). Dual labor market and the "Phillips Curve Puzzle": The Japanese experience. *Journal of Evolutionary Economics*.

Aoyama, H., Fujiwara, Y., Ikeda, Y., Soma, W., & Yoshikawa, H. (2017). *Macro-econophysics-new studies on economic networks and synchronization*. Cambridge University Press.

Aoyama, H., Iyetomi, H., & Yoshikawa, H. (2015). Equilibrium distribution of labor productivity: a theoretical model. *Journal of Economic Interaction and Coordination, 10*(1), 57–66. (April).

Arrow, K. (1959). Toward a theory of price adjustment. In M. Abramovitz (Ed.), *The allocation of economic resources*. Stanford University Press.

Arrow, K. (1963). The role securities in the optimal allocation of risk-bearing. *Review of Economic Studies, 31*, 91–96.

Arrow, K. (1967). Samuelson collected. *Journal of Political Economy, 75*(5), 730–737.

Arrow, K., & Debreu, G. (1954). Existence of an equilibrium for a competitive economy. *Econometrica, 22*(3), 265–290.

Attanasio, O. P. (1999). Consumption. In *Handbook of Macroeconomics* (Vol. 1, pp. 741–812).

Auerbach, A. J., & Obstfeld, M. (2005). The case for open-market purchases in a liquidity trap. *American Economic Review, 95*(1), 110–137.

Bak, P., Chen, K., Scheinkman, J., & Woodford, M. (1993). Aggregate fluctuations from independent sectoral shocks: Self-organized criticality in a model of production and inventory dynamics. *Ricerche Economiche, 47*, 3–30.

Ball, L. (2006). Is the 'accounting' theory of inflation wrong?. *NBER Working Paper* No. 12687.

Ball, L., Mankiw, N. G., & Romer, D. (1988). The new Keynesian economics and the output-inflation tradeoff. *Brookings Paper on Economic Activity, 1*, 1–65.

Bando, M. (2009). From traffic flow to economic system. *Progress of Theoretical Physics Supplement*, No. 179, 189–197.

Barabási, A. (2016). *Network science*. Cambridge University Press.

Barro, R., & Becker, G. (1989). Fertility choice in a model of economic growth. *Econometrica, 57*(2), 481–501.

Bass, F. (1969). A new product growth for model consumer durables. *Management Science, 15*(5), 215–227.

Basu, S. (1996). Procyclical productivity: Increasing returns or cyclical utilization? *Quarterly Journal of Economics, 111*(3), 719–751.

Becker, G. (1981). *A treatise on the family*. Harvard University Press.

Benassy, J. P. (1975). Neo-Keynesian disequilibrium theory in a monetary economy. *The Review of Economic Studies, 42*(4), 503–523.

Bernanke, B. S. (2000). Japanese monetary policy: A case of self-induced paralysis. In R. Mikitani, & A. S. Posen (Eds.), *Japan's financial crisis and its parallels to U.S. experience* (pp. 149–166). Institute for International Economics.

Bernanke, B. S. (2020). The new tools of monetary policy. *American Economic Review, 110*(4), 943–983.

Bernanke, B. S., Reinhart, V. R., & Sack, B. S. (2004). Monetary policy alternatives at the zero bound: An empirical assessment. *Brooking Papers on Economic Activity, 2*, 1–100.

Bils, M., & Klenow, P. J. (2004). Some evidence on the importance of sticky prices. *Journal of Political Economy, 112*, 947–985.

Black, F. (1986). Noise. *Journal of Finance, 41*(3), 529–543.

Black, F. (1995). *Exploring general equilibrium*. MIT Press.

Blanchard, O. (1979). Speculative bubbles, crashes and rational expectations. *Economics Letters, 3*, 387–389.

Blanchard, O. (2018). Should we reject the natural rate hypothesis? *Journal of Economic Perspectives, 32*(1), 97–120.

Blanchard, O., & Fischer, S. (1989). *Lectures on macroeconomics*. MIT Press.

Blanchard, O., & Watson, M. (1982). Bubbles, rational expectations and financial markets. In P. Wachtel (Ed.), *Crises in the economic and financial structure* (pp. 295–316). D. C. Heathand Company.

Blinder, A., & Maccini, L. (1991). Taking stock: A critical assessment of recent research on inventories. *Journal of Economic Perspectives, 5*(1), 73–96.

Bloom, N., Jones, C. I., van Reenen, J., & Webb, M. (2020). Are ideas getting harder to find? *American Economic Review, 110*(4), 1104–1144.

Borio C., Erdem, M., Filardo, A., & Hofmann. B. (2015). The costs of deflations: a historical perspective. *BIS Quarterly Review*, 31–54. (March).

Bouchaud, J., & Potters, M. (2003). *Theory of financial risk and derivative pricing: From statistical physics to risk management* (2nd ed.). Cambridge University Press.

Boushey, H., DeLong, J. B., & Steinbaum, M. (Eds.). (2017). *After Piketty: The Agenda for economics and inequality*. Harvard University Press.

Brentano, L. (1910). The doctrine of malthus and the increase of population during the last decades. *Economic Journal, 20*(79), 371–393.

Brock, W. A. (1999). Scaling in economics: A reader's guide. *Industrial and Corporate Change, 8*(3), 409–446.

Brunnermeier, M. K. (2009). Deciphering the liquidity and credit crunch 2007–2008. *Journal of Economic Perspectives, 23*(1), 77–100.

Bruno, M., & Sachs, J. (1985). *Economics of worldwide stagflation*. Harvard University Press.

Brynjolfsson, E., & Hitt, L. (2000). Beyond computation: Information technology, organizational transformation and business performance. *Journal of Economic Perspective, 14*(4), 23–48.

Burdett, K., & Mortensen, D. T. (1998). Wage differentials, employer size, and unemployment. *International Economic Review, 39*(2), 257–273.

Caballero, R. J. (1999). Aggregate investment. *Handbook of Macroeconomics, 1*, 813–862.

Caballero, R. J. (2010). Macroeconomics after the crisis: Time to deal with the pretense-of-knowledge syndrome. *Journal of Economic Perspectives, 24*(4), 85–102.

Calvo, G. A. (1983). Staggered prices in a utility-maximizing framework. *Journal of Monetary Economics, 12*(3), 383–398.

Campbell, J., & Cochrane, J. (1999). By force of habit: A consumption-based explanation of aggregate stock market behavior. *Journal of Political Economy, 107*, 205–251.

Channing, D., Amartel, L. A. N., Lee, Y., Meyer, M., & Stanley, H. E. (1998). Scaling the volatility of GDP growth rates. *Economic Letters, 60*, 335–341.

Cassel, G. (1922). *Money and foreign exchange after 1914*. Oxford University Press.

Cassel, G. (1932). *The theory of social economy*. Harcourt, Brace & Company.

Cassidy, J. (2010). Interview with Eugene Fama. *New Yorker*, November, 1.

Cecchetti, S., Lam, P., & Mark, N. (2000). Asset pricing with distorted beliefs: Are equity returns too good to be true? *American Economic Review, 90*(4), 787–805.

Champernowne, D. (1953). A model of income distribution. *Economic Journal, 83*, 318–351.

Champernowne, D. (1973). *The distribution of income between persons*. Cambridge University Press.

Clay, H. (1928). Unemployment and wage rates. *Economic Journal, 38*(149), 1–15.

Clower, R. (1965). The Keynesian counterrevolution: A theoretical appraisal. In F. Hahn, & F. Brechling (Eds.), *The theory of interest rates*. Macmillan.

Costello, D. (1993). A cross-country, cross-industry comparison of productivity growth. *Journal of Political Economy, 101*(2), 207–222.
Coutts, K., Godley, W., & Nordhaus, W. (1978). *Industrial priming in the United Kingdom.* Cambridge University Press.
Davis, S. J., Haltiwanger, J. C., & Schuh, S. (1996). *Job creation and destruction.* MIT Press.
Davis, S. J., Faberman, R. J., & Haltiwanger, J. C. (2013). The Establishment-level behavior of vacancies and hiring. *Quarterly Journal of Economics, 128*(2), 581–622.
Deaton, A. (1992). *Understanding consumption.* Oxford University Press.
Debreu, G. (1959). *Theory of value.* Wiley.
Delli Gatti, D., Gallegatti, G., Giulioni, G., Palestrini, A. (2008). *Emergent macroeconomics: An agent-based approach to business fluctuations.* Springer
Diamond, P. (1967). The role of a stock market in a general equilibrium model. *American Economic Review, 57*(4), 759–776.
Diamond, P. (1982). Aggregate demand management in search equilibrium. *Journal of Political Economy, 90*(5), 881–894.
Diamond, P. (2011). Unemployment, vacancies, and wages. *American Economic Review, 101*(4), 1045–1072.
Diebold, F., & Rudebusch, G. (1990). A nonparametric investigation of duration dependence in the American business cycle. *Journal of Political Economy, 98*, 596–616.
Di Guilmini, C. Landini, & S. Gallegatti, M. (2017). *Interactive macroeconomics: Stochastic aggregate dynamics with heterogenerous and interacting agents.* Cambridge University Press.
Di Matteo, M., & Yoshikawa, H. (1999). Economic growth: The Italian and Japanese experiences. *Economic Systems, 23*(11).
Domer, E. (1946). Capital expansion, rate of growth and employment. *Econometrica, 14*, 137–147.
Drèze, J. (1979). Demand estimation, risk-aversion, and sticky prices. *Economic Letters, 4*, 1–6.
Drèze, J., & Herings, P. J. (2008). Kinky perceived demand curves and Keynes-Negishi equilibria. *International Journal of Economic Theory, 4*, 207–246.
Dunlop, J. T. (1948). Productivity and the wage structure. In L. Metzler (Ed.), *Income, employment and public policy: Essays in honor of Alvin H. Hansen.* Norton.
Durlauf, S. N. (1997). Statistical mechanics approaches to socioeconomic behavior. In B. Arthur, S. N. Durlauf, & Lane (Eds.), *The economy as an evolving complex system II.* Addison-Wesley.
Economist. (2009). Where modern economic theory went wrong and how the crisis is changing it. *Economist*, July 18th–24th .
Eggertsson, G.B., Woodford, M. (2003). The zero bound on interest rates and optimal monetary policy, Brookings Papers on Economic Activity No. 1, 139–233.
Fair, R. C. (1989). Book review of R. E. Lucas, models of business cycles, London and New York: Blackwell. *Journal of Economic Literature, 27*, 104–105.
Fama, E. (1965). Portfolio analysis in a stable Paretian market. *Management Science, 11*, 404–419.
Fama, E. (1970). Efficient capital markets: A review of theory and empirical work. *The Journal of Finance, 25*(2), 383–417.
Fay, J. A., & Medoff, J. L. (1985). Labor and output over the business cycle: Some direct evidence. *American Economic Review, 75*(4), 638–655.
Feller, W. (1971). *An introduction to probability theory and its applications* (2nd ed., Vol. 2). Wiley.
Feynman, R. P. (1964). The principle of least action. *Lectures on Physics Reading.* Addison Wesley, Chapter 19.
Fisher, I. (1933). The debt-deflation theory of Great Depression. *Econometrica, 1*(4), 337–357.
Fisher, J. C., & Pry, R. H. (1971). A simple substitution model of technological change. *Technological Forecasting and Social Change, 3*, 75–88.
Foley, D. (1994). A statistical equilibrium theory of markets. *Journal of Economic Theory, 62*, 321–345.
Frankel, J., & Froot, K. (1987). Short-term and long-term expectations the Yen/Dollar exchange rate: Evidence brow survey data. *Journal of the Japanese and International Economies, 1*(3), 249–274.

Freeman, R. B., & Weitzman, M. (1987). Bonuses and employment in Japan. *Journal of the Japanese and International Economies, 1*, 168–194.
Freeman, R. B., & Rebick, M. E. (1989). Crumbling pillar? Declining union density in Japan. *Journal of the Japanese and International Economies, 3*, 578–605.
Friedman, M. (1957). *A theory of the consumption function.* Princeton University Press.
Friedman, M. (1968). The role of monetary policy. *American Economic Review, 58*(1), 1–17.
Friedman, M. (1974). Perspective on inflation (p. 73). *Newsweek.* (June 24).
Friedman, M. (1987). Quantity theory of money. In J. Eatwell, M. Milgate, & P. Newman (Eds.), *The New Palgrave: A dictionary of economics.* Macmillan.
Frisch, R. (1931). A method of decomposing an empirical series into its cyclical and progressive components. *Journal of American Statistical Association, 26*, 78.
Frisch, R. (1933). Propagation problems and impulse problems in dynamic economics. *Economic Essays in Honour of Gustav Cassel.* Allen & Unwin.
Fujiwara, Y. (2020). Firms growth, distribution, and non-self-averaging revisited. In H. Aoyama, Y. Aruka, & H. Yoshikawa (Eds.), *Complexity, heterogeneity, and the methods of statistical physics in economics* (pp. 117–141). Springer.
Fukao, K., Nakamura, N., & Nakabayashi, S. (2017a, b, 2018), *Economic History of Japan*, (in Japanese) (Vol. 3, 4, 5). Iwanami Shoten.
Fukao, T. (1987). *Decentralized system: Thermodynamic approach.* (in Japanese) Shokodo
Gabaix, X. (2009). Power laws in economics and finance. *Annual Review of Economics, 1*, 255–293.
Gabaix, X. (2011). The granular origins of aggregate fluctuations. *Econometrica, 79*(3), 733–772.
Gabaix, X., Gopikrishnan, P., Plerou, V., & Stanley, H. E. (2003). A theory of power-law distributions in financial market fluctuations. *Nature, 423*, 267–270.
Jordi, G. (2011). *Unemployment fluctuations and stabilization policies.* MIT Press.
Garibaldi, U., & Scalas, E. (2010). *Finitary probabilistic methods in econophysics.* Cambridge University Press.
Gennaioli, N., & Shleifer, A. (2018). *A crisis of belief*. Prinston University Press.
Gibrat, R. (1931). *Les Inégalités Économiques.* Recueil Sirey.
Gnedenko, B., & Kolmogorov, A. N. (1954). *Limit distributions for sums of independent random variables.* Addison-Wesley.
Gnedenko, B., & Ushakov, I. A. trans. (1997). *Theory of probability* (6th ed.) (English translation of the Russian original published in 1988). Gordon and Breach Science Publishers.
Godley, W., & Nordhaus, W. (1972). Pricing in the trade cycles. *Economic Journal, 82*(327), 853–882.
Goldie, C. M. (1991). Implicit renewal theory and tails of solutions of random equations. *The Annals of Applied Probability, 1*, 126–166.
Goldsmith, R. W. (1969). *Financial structure and development.* Yale University Press.
Golosov, M., & Lucas, R. (2007). Menu costs and phillips curves. *Journal of Political Economy, 115*, 171–199.
Gordon, R. J. (1982). Why U. S. wages and employment behaviuor differs from that in Britain and Japan. *Economic Journal, 92*, 13–44.
Gordon, R. J. (2011). The history of the Phillips Curve: Consensus and bifurcation. *Economica, 78*(309), 10–50.
Gordon, R. J. (2016). *The rise and fall of American growth.* Princeton University Press.
Granger, C. (1966). The typical spectral shape of an economic variable. *Econometrica, 34*(1), 150–161.
Greenspan, A. (1996). Achieving price stability. *Introductory Comments for a Symposium Sponsored by the Federal Reserve Bank of Kansas City.*
Griliches, Z. (1957). Hybrid corn: An exploration in the economics of technological change. *Econometrica, 25*(4), 501–522.
Grossman, G. M., & Helpman, E. (1991). *Innovation and growth in the global economy.* MIT Press.
Grossman, S., & Shiller, R. (1981). The determinants of the variability of stock market prices. *American Economic Review, 71*(2), 222–227.

Haberler, G. (1964). *Prosperity and depression* (5th ed.). Harvard University Press.
Hagedorn, M., & Manovskii, I. (2008). The cyclical behavior of equilibrium unemployment and vacancies revisited. *American Economic Review, 98*(4), 1692–1706.
Hahn, F., & Solow, R. (1995). *A critical essay on modern macroeconomic theory*. MIT Press.
Haken, H. (1977). *Synergetics*. Springer.
Hall, B. H. (2011). Innovation and productivity. *NBER Working Paper*, No. 17178.
Hall, R. (1978). Stochastic implications of the life cycles permanent income hypothesis: Theory and evidence. *Journal of Political Economy, 86*(6), 971–987.
Hall, R., & Sargent, T. J. (2018). Short-run and long-run effects of Milton Friedman's presidential address. *Journal of Economic Perspectives, 32*(1), 121–134.
Hall, R. L., Hitch, C. J. (1939). Price theory and business behavior. *Oxford Economic Papers*, No. 2, 12–45.
Haltiwanger, J., Hyatt, H., Khan, L., & McEntarfer, E. (2018). Cyclical job ladders by firm size and firm wage. *American Economic Journal Macroeconomics, 10*(2), 52–85.
Hansen, A. (1939). Economic progress and declining population growth. *American Economic Review, 29*(1), 1–15.
Harrod, R. (1939). An essay in dynamic theory. *Economic Journal, 49*(193), 14–33.
Hayashi, F. (1982). Tobin's marginal q and average q: A neoclassical interpretation. *Econometrica, 50*(1), 213–224.
Hayashi, F., & Prescott, E. C. (2002). The 1990s in Japan: A lost decade. *Review of Economic Dynamics, 5*(1), 206–235.
Hicks, J. R. (1936). Mr. Keynes's theory of employment. *Economic Journal, 46*(182), 238–253.
Hicks, J. R. (1937). Mr. Keynes and the classics: A suggested interpretation. *Econometrica, 5*(2), 147–159.
Hicks, J. R. (1946). *Value and capital* (2nd ed.). Oxford University Press.
Hicks, J. R. (1950). *A contribution to the theory of the trade cycle*. Oxford University Press.
Hicks, J. R. (1973). The mainspring of economics growth. *Swedish Journal of Economics, 75*, 336–348. (also In id. *Economic perspectives: Further essay on money and growth*, Oxford: Oxford University Press, 1977).
Hicks, J. R. (1974a). *The Crisis in Keynesian Economics*. Basic Books.
Hicks, J. R. (1974b). Real and monetary factors in economic fluctuations. *Scottish Journal of Political Economy, 21*(3), 205–214.
Hicks, J. R. (1977). Monetary experience and the theory of money. In id. *Economic perspectives: Further essay on money and growth*. Oxford University Press.
Hicks, J. R. (1989). *A market theory of money*. Oxford University Press.
Honig, J. M. (2009). The role of Fisher information theory in the development of fundamental laws in physical chemistry. *Journal of Chemical Education, 86*(1), 116–119.
Hornstein, A. (1998). Inventory investment and the business cycle. *Fed Reserve Bank Richmond Economic Quarterly, 84*, 49–71.
Houthakker, H. (1955/56). The Pareto distribution and the Cobb-Douglas production function in activity analysis. *Review of Economic Studies, 23*(1), 27–31.
Huang, Z. F., & Solomon, S. (2001). Power, levy, exponential and Gaussian-like regimes in aut6catalytic financial systems. *The European Physical Journal B: Condensed Matter and Complex Systems, 20*, 601–607.
Ijiri, Y., & Simon, H. A. (1975). Some distributions associated with Bose-Einstein Statistics. *Proceedings of National Academy of Sciences, U.S.A., 72*(5), 1654–1657.
Ijiri, Y., & Simon, H. A. (1979). *Skew distributions and the sizes of business firms*. North-Holland Pub. Co.
Ikeda, Y., & Souma, W. (2009). International comparison of labor productivity distribution for manufacturing and non-manufacturing firms. *Progress of Theoretical Physics Supplement, 179*, 93–102.
Iwai, K. (1981). *Disequilibrium dynamics: A theoretical analysis of inflation and unemployment*. Yale University Press.

Iyetomi, H. (2012). Labor productivity distribution with negative temperature. *Progress of Theoretical Physics, Supplement*, No. 194, 135–143.

Iyetomi, H., Nakayama, Y., Yoshikawa, H., Aoyama, H., Fujiwara, Y., Ikeda, Y., & Souma, W. (2011). What causes business cycles?: Analysis of Japanese industrial production data. *Journal of Japanese and International Economies, 25*(3), 246–272.

Javanovic, B. (1987). Micro shocks and aggregate risk. *Quarterly Journal of Economics, 102*(2), 395–410.

Johansen, A., & Sornette, D. (1998). Stock market crashes are outliers. *The European Physical Journal B, 1*, 141–143.

Jorgenson, D., Ho, M., & Stiroh, K. (2005). *Productivity Vol. 3: Information Technology and the American Growth Resurgence*. MIT Press.

Kadanoff, L. (2000). *Statistical physics*. World Scientific Publishing.

Kahneman, D., Knetsch, J., & Thaler, R. (1986). Fairness as a constraint on profit seeking: Entitlement in the market. *American Economic Review, 76*(4), 728–741.

Kaldor, N. (1957). A model of economic growth. *Economic Journal, 68*, 591–624.

Kaldor, N. (1966). *Causes of the slow rate of economic growth of the United Kingdom: An inaugural lecture*. Cambridge University Press.

Kaldor, N. (1970). The new monetarism. *Llydos Bank Review*, July.

Kaldor, N. (1982). *The scourge of monetarism*. Oxford University Press.

Kaldor, N. (1985). *Economics without equilibrium*. University of Cardiff Press.

Kalecki, M. (1954). *Theory of economic dynamics*. George Allen & Unwin.

Kesten, H. (1973). Random difference equations and renewal theory for products of random matrices. *Acta Mathematica, 131*, 207–248.

Keynes, J. M. (1919). *The economic consequences of the peace*. Macmillan. (Collected Writings of J. M. Keynes, Vol.2).

Keynes, J. M. (1923). *A tract on monetary reform*. Macmillan. (Collected Writings of J. M. Keynes, Vol. 4).

Keynes, J. M. (1924). "Alfred Marshall," Economic Journal, September, also in *Essays in Biography*. Macmillan. (Collected Writings of J. M. Keynes, Vol. 10).

Keynes, J. M. (1930), *A treatise on money* (Vols. 1 & 2). Macmillan. (Collected Writings of J. M. Keynes, Vols. 5 and 6).

Keynes, J. M. (1931). The consequences to the banks of the collapse of money values (August). In id. *Essays in persuasion*. Macmillan. (Collected Writings of J. M. Keynes, Vol. 9).

Keynes, J. M. (1936). *The general theory of employment, interest, and money*. Macmillan.

Keynes, J. M. (1937). The general theory of employment. *Quarterly Journal of Economics, 51*(2), 209–223. (Feb.), (Collected Writings of J. M. Keynes, Vol. 14).

Keynes, J. M. (1973), *The general theory and after, Part II defense and development*. (Collected Writings of J. M. Keynes, Vol. 14).

Kichikawa, Y., Aoyama, H., Fujiwara, Y., Iyetomi, H., Yoshikawa, H. (2018). Analysis of inflation/deflation: Clusters of micro prices matter!. *RIETI Discussion Paper Series*, 18-E-055.

Kichikawa, Y., Iyetomi, H., Aoyama, H., Fujiwara, Y., & Yoshikawa, H. (2020). Interindustry linkages of prices: Analysis of Japan's deflation. *PLoS ONE*. https://doi.org/10.1371/journal.pone.0228026

Kikuchi, S., Kitao, S., & Mikoshiba, M. (2021). Who suffers from the COVID-19 shocks? Labor market heterogeneity and welfare consequences in Japan. *Journal of The Japanese and International Economies, 59*, 101117.

Kindleberger, C. (1967). *Europe's postwar growth*. Harvard University Press.

Kindleberger, C. (1978). *Manias, panics and crashes: A history of financial crises*. Macmillan.

King, M. (2016). *The end of alchemy: money, banking, and the future of the global economy*. W. W. Norton.

Kirman, A. (1992). Whom or what does the representative individual represent? *Journal of Economic Perspectives, 6*(2), 117–136.

Kittel, C., & Kroemer, H. (1980). *Thermal physics* (2nd ed.). W. H. Freeman.

Kiyotaki, N. (2011). A perspective on modern business cycle theory. *Economic Quarterly, 97*(3), 195–208.
Kiyotaki, N., & Moore, J. (1997). Credit cycles. *Journal of Political Economy, 105*(2), 211–248.
Klein, J. (1999). The rise of 'Neo-October' econometrics: Kondratiev and Slutzky at Moscow Conjuncture Institute. *History of Political Economy, 31*(4), 152.
Klenow, P. J., Malin, B. A. (2011). Microeconomic evidence on price-setting. In B. H. Friedman & M. Woodford (Eds.), *Handbook of monetary economics* (Vol. 3A), Chapter 6. North-Holland.
Kondratiev, N. D. (1926). The problem of prediction. In N. Makasheva, W. Samuels, & V. Barnett (Eds.), *The works of Nikolai D. Kondratiev* (Vol. 3). Pickering and Chatto 1998.
Krugman, P. (1991). History versus expectations. *Quarterly Journal of Economics, 106*(2), 651–667.
Krugman, P. (1994). The myth of Asia's miracle. *Foreign Affairs*, November/December.
Krugman, P. (1998). It's Baaack: Japan's slump and the return of the liquidity trap. *Brookings Papers on Economics Activity, 2*, 137–203.
Krusell, P., & Smith, A. (1998). Income and wealth heterogeneity in the macroeconomy. *Journal of Political Economy, 106*(5), 867–896.
Kubo, R. (1952). *Statistical mechanics* (in Japanese). Kyoritsu.
Kuznetz, S. (1929). Random events and cyclical oscillations. *Journal of American Statistical Association, 24*(167), 273.
Kuznetz, S. (1940). Schumpeter's business cycles. *American Economic Review, 30*, 257–271. (June).
Kuznetz, S. (1953). *Economic Change: Selected essays in business cycles, national income, and economic growth,* W. W. Norton.
Kydland, F., & Prescott, E. (1982). Time to build and aggregate fluctuation. *Econometrica, 50*(6), 1345–1370.
Leiyonhufvud, A. (1968). *Keynesian economics and economics of Keynes*. Oxford University Press.
Leontief, W. (1936). The fundamental assumption of Mr. Keynes's monetary theory of unemployment. *Quarterly Journal of Economics, 51*(1), 192–197. (November).
Lester, R. (1946). Shortcomings of marginal analysis for wage-employment problems. *American Economic Review, 36*(1), 63–82.
Lévy, M., & Solomon, S. (1996). Power laws are logarithmic Boltzmann laws. 338 *International Journal of Modern Physics, C, 7*(4), 595–601.
Lewis, A. (1954). Economic development with unlimited supplies of labour. *Manchester School of Economic and Social Studies, 22*(2), 139–191.
Lo, A. W. (2017). *Adaptive markets*. Princeton University Press.
Lucas, R. E. (1972). Expectation and the neutrality of money. *Journal of Economic Theory, 4*(2), 103–124.
Lucas, R. E. (1973). Some international evidence on output-inflation tradeoffs. *American Economic Review, 63*(3), 326–334.
Lucas, R. E. (1976). Economic policy evaluation: A critique. *Carnegie-Rochester Conference Series on Public Policy, 1*, 19–46.
Lucas, R. E. (1987). *Models of business cycles*. Blackwell.
Lucas, R. E. (1988). On the mechanics of economic development. *Journal of Monetary Economics, 22*, 3–42. (July).
Lucas, R. E. (2003). Macroeconomic priorities. *American Economic Review, 93*(1), 1–14.
Lucas, R. E., & Prescott, E. (1974). Equilibrium search and unemployment. *Journal of Economic Theory, 7*(2), 721–754.
Lux, T., & Sornette, D. (2002). On rational bubbles and fat tails. *Journal of Money, Credit and Banking Part* 1, *34*(3), 589–610.
Malevergne, Y., & Sornette, D. (2006). *Extreme financial risks*. Springer.
Malinvaud, E. (1977). *The theory of unemployment reconsidered*. Basil Blackwell.
Malinvaud, E. (1982). Wages and unemployment. *Economic Journal, 92*(365), 1–12.
Malinvaud, E. (1994). *Diagnosing unemployment*. Cambridge University Press.
Mandelbrot, B. (1963a). The variation of certain speculative prices. *Journal of Business, 36*(4), 394–419.

Mandelbrot, B. (1963b). New methods in statistical economics. *Journal of Political Economy, 71*(5), 421–440.
Mandelbrot, B. (1997). *Fractals and scaling in finance*. Springer.
Mandelbrot, B., & Hudson, R. L. (2004). *The (MIS) behavior of markets*. Basic Books.
Mankiw, N. G. (1985). Small menu costs and large business cycles: A macroeconomic model of monopoly. *Quarterly Journal of Economics, 100*(2), 529–537.
Mankiw, N. G. (1989). Real business cycles: A new Keynesian perspective. *Journal of Economic Perspectives, 3*(3), 79–90.
Mankiw, N. G. (2016). *Macroeconomics* (9th ed.). Worth Publishers.
Mankiw, N. G., & Reis, R. (2018). Friedman's presidential address in the evolution of macroeconomic thought. *Journal of Economic Perspectives, 32*(1), 81–96.
Mankiw, N. G., & Romer, D. (1991). *New Keynesian economics* (Vols. 1 and 2). The MIT Press.
Mantegna, R., & Stanley, H. E. (2000). *An Introduction to econophysics: Correlations and complexity in finance*. Cambridge University Press.
Marshall, A. (1587). Remedies for fluctuations of general prices. *Contemporary Review*. (March, also in A. C. Pigou ed., *Memorials of Alfred Marshall*. MacMillan, 1925).
Marshall, A. (1898). Mechanical and biological analogies in economics. (also in A. C. Pigou (Ed.), *Memorials of Alfred Marshall*. Macmillan, 1925).
Marshall, A. (1920). *Principles of economics* (8th ed.). Macmillan.
Marshall, A. (1926). *Official papers by Alfred Marshall*. In J. M. Keynes (Ed.). MacMillan.
Marsili, M., & Zhang, Y. (1998). Interacting individuals leading to Zipf's law. *Physical Review Letters, 80*(12), 1073–1091.
Murayama, T. (2008). *Portrait of Walras* (in Japanese). Keiso Shobo.
McCauley, J. (2004). *Dynamics of markets: Econophysics and finance*. Cambridge University Press.
McLaughlin, G. & Watkins, R. (1939). The problem of industrial growth in a mature economy. *American Economic Review, 29*(1), 1–14.
Mehra, R., & Prescott, E. (1985). The equity premium. *Journal of Monetary Economics, 15*(2), 145–161.
Merton, R. (1973). An intertemporal capital asset pricing model. *Econometrica, 41*(5), 867–887.
Metzler, L. (1941). The nature and stability of inventory cycles. *Review of Economics and Statistics, 23*(3), 113–129.
Midrigan, V. (2011). Menu costs, multiproduct firms, and aggregate fluctuations. *Econometrica, 79*(4), 1139–1180.
Minsky, H. (1975). *John Maynard Keynes*. Columbia University Press.
Minsky, H. (1986). *Stabilizing an unstable economy*. Yale University Press. (New York: McGraw-Hill, 2008).
Modigliani, F. (1944). Liquidity preference and the theory of interest and money. *Econometrica, 12*(1), 45–88.
Modigliani, F. (1986). Life cycle, individual thrift, and the wealth of nations. *American Economic Review, 76*, 297–313.
Modigliani, F., & Ando, A. (1963). The 'life-cycle' hypothesis of saving: Aggregate implications and tests. *American Economic Review, 53*, 55–84.
Modigliani, F., & Brumberg, R. (1954). Utility analysis and the consumption function: An interpretation of cross-section data. In K. Kurihara, (Ed.), *Post-Keynesian economics*. Rutgers University Press.
Montroll, E. (1981). On the entropy function in sociotechnical systems. In *Proceedings of National Academy of Sciences, U.S.A.*, Vol. 78, No. 12, pp. 7839–7893.
Montroll, E. (1987). On the dynamics and evolution of some socio-technical systems. *Bulletin (New Series) of American Mathematical Society, 16*(1), 1–46.
Mortensen, D. T. (2003). *Wage dispersion*. MIT Press.
Mortensen, D. T. (2011). Markets with search friction and the DMP model. *American Economic Review, 101*(4), 1073–1091.

Mortensen, D. T., & Pissarides, C. (1994). Job creation and job destruction in the theory of unemployment. *The Review of Economic Studies, 61*(3), 397–415.
Motonishi, T., & Yoshikawa, H. (1999). Causes of the long stagnation of Japan during the 1990's: Financial or real? *Journal of the Japanese and International Economies, 13*(3), 181–200.
Mueller, A. I. (2017). Separations, sorting, and cyclical unemployment. *American Economic Review, 107*(7), 2081–2107.
Muth, J. F. (1960). Optimal properties of exponentially weighted forecasts. *Journal of American Statistical Association, 55*(290).
Muth, J. F. (1961). Rational expectations and the theory of price movements. *Econometrica, 29*(3), 384–404.
Myrdal, G. (1933). Industrialization and population. In *Economic essays in honor of Gustav Cassel*. George Allen and Unwin.
Neftci, S. (1984). Are economic time series asymmetrical over the business cycles? *Journal of Political Economy, 92*(2), 307–328.
Negishi, T. (1960). Welfare economics and existence of an equilibrium for a competitive economy. *Metroeconomica, 12*(2–3), 92–97.
Negishi, T. (1960–61). Monopolistic competition and general equilibrium. *Review of Economic Studies, 28*(3), 196–201.
Negishi, T. (1965). *Theory of Price and Allocation* (in Japanese). Toyokeizai.
Negishi, T. (1979). *Microeconomic foundations of Keynesian macroeconomics*. North-Holland.
Newman, M. (2005). Power laws, Pareto distributions and Zipf's law. *Contemporary Physics, 46*(5), 323–351.
Nirei, M. (2006). Threshold behavior and aggregate fluctuation. *Journal of Economic Theory, 127*(1), 309–322.
Nirei, M., & Souma, W. (2007). Two factor model of income distribution dynamics. *Journal of Income and Wealth, 53*(3), 440–459.
Ohkawa, K., & Rosovsky, H. (1973). *Japanese economic growth*. Stan-ford University Press.
Okada, M. (2006). *Economic Dynamics of Kondratiev: long cycles and probabilistic philosophy*. (in Japanese) Sekaishoin
Okun, A. M. (1962). Potential GNP: Its measurement and significance. In *Proceedings of the Business and Economic Statistics Section of the American Statistical Association* (pp. 98–104). (reprinted in id., *The Political Economy of Prosperity*, Washington D.C.: The Brookings Institution, 1970).
Okun, A. M. (1973). Upward mobility in a high-pressure economy. *Brookings Papers on Economic Activity* (1).
Okun, A. M. (1981). *Prices and quantities: A macroeconomic analysis*. The Brookings Institution.
Pareto, V. (1896). *Cour d'Economie Politique*. Lausanne.
Pasinetti, L. (1981). *Structural change and economic growth*. Cambridge University Press.
Pasinetti, L. (1993). *Structural economic dynamics: A theory of the economic consequences of human learning*. Cambridge University Press.
Peneva, E., & Rudd, J. (2017). The passthrough of labor costs to price inflation. *Journal of Money, Credit and Banking, 49*(8), 1777–1802.
Phelps, E. (1970). *Microeconomic foundations of employment and inflation theory*. W.W. Norton.
Phelps Brown, H., & Hart, P. (1952). The share of wages in national income. *Economic Journal, 62*(246), 253–277. (June).
Phelps Brown, H., & Ozga, S. A. (1955). Economic growth and the price level. *Economic Journal, 65*, 1–18.
Phillips, A. (1958). The relation between unemployment and the rate of change of money wages in the United Kingdom, 1861–1957. *Economica, 25*(100), 283–299.
Pigou, A. (1925). *Memorials of Alfred Marshall*. Macmillan.
Pigou, A. (1927). Wage policy and unemployment. *Economic Journal, 37*(147), 355–368.
Pigou, A. (1929). *Industrial fluctuations* (2nd ed.). Macmillan.
Pigou, A. (1933). *The theory of unemployment*. Macmillan.

Pigou, A. (1935). Inflation, deflation and reflation. In id., *Economics in practice* (pp. 80–106). MacMillan.
Pigou, A. (1936). Mr. J. M. Keynes' general theory of employment, interest and money. *Economica, 3*(10), 115–132.
Pigou, A. (1950). *Keynes's "General Theory": A retrospective view*. Macmillan.
Piketty, T. (2014), *Capital in the twenty-first century*. Harvard University Press.
Pissarides, C. A. (2000). *Equilibrium unemployment theory* (2nd ed.). MIT Press.
Pissarides, C. A. (2009). The unemployment volatility puzzle: Is wage stickiness the answer? *Econometrica, 77*(5), 1339–1369.
Pissarides, C. A. (2011). Equilibrium in the labor market search frictions. *American Economic Review, 101*(4), 1092–1105.
Postan, M. M. (1971). Technological progress in post-war Europe. In id., *Fact and relevance*, Chapter 11. Cambridge University Press.
Postel-Vinay, F., Robin, J. (2002). Equilibrium wage dispersion with worker and employer heterogeneity. *Econometrica, 70*(6), 2295–2350.
Prescott, E. (1998). Needed: A theory of total factor productivity. *International Economic Review, 39*(3), 525–551.
Quesnay, F. (1758), Tableau Économique.
Ramsey, F. (1928). A mathematical theory of saving. *Economic Journal, 38*(152), 543–559.
Ricardo, D. (1817). Principles of political economy, and taxation. London. (3rd ed. 1821, Vol. 1 of P. Sraffa ed. *Works and Correspondence of David Ricardo*, Cambridge (UK): Cambridge University Press, 1953).
Redner, S. (1990). Random multiplicative processes: An elementary tutorial. *American Journal of Physics, 58*(3), 267–273.
Robertson, D. H. (1915). *A study of industrial fluctuation*. P. S. King & Son. (reprinted with new Introduction by London School of Economics and Political Science, 1948).
Robertson, D. H. (1931). The world slump. In A. C. Pigou, D. H. Robertson (Eds.), *Economic essays and addresses*. P. S. King & Son.
Robertson, D. H. (1936). Some notes on Mr. Keynes' general theory of employment. *Quarterly Journal of Economics, 51*(1), 168–170, 187–191. (September and November).
Robinson, J. (1962). *Essays in the theory of economic growth*. Macmillan.
Roehner, B. (2009). *Hidden collective factors in speculative trading: A study in analytical economics* (2nd ed.). Springer.
Rogerson, R., Shimer, R., & Wright, R. (2005). Search theoretic models of the market: A survey. *Journal of Economic Literature, 43*(4), 959–988.
Romer, P. M. (1986). Increasing returns and long run growth. *Journal of Political Economy, 94*(5), 1002–1037.
Rostow, W. W. (1948). *British economy of the nineteenth century*. Ox-ford University Press.
Rostow, W. W. (1953). *The process of economic growth*. Oxford University Press.
Rostow, W. W. (1960). *The stages of economic growth: A non-communist manifest*. Cambridge University Press.
Rostow, W. W. (1978). *The world economy*. University of Texas Press.
Saichev, A., Malevergne, Y., & Sornette, D. (2010). *Theory of Zipf 's law and beyond*. Springer.
Salter, W. E. G. (1960). *Productivity and technical change*. Cambridge University Press.
Samorodnitsky, G., & Taqqu, M. S. (1994). *Stable non Gaussian random process: Stochastic models with infinite variance*. Chapman Hall.
Samuelson, P. A. (1939). Interactions between the multiplier analysis and the principle of Acceleration. *Review of Economic Statistics, 21*(2), 75–78.
Samuelson, P. A. (1946). Lord Keynes and the general theory. *Econometrica, 14*(3), 187–200.
Samuelson, P. A. (1948). *Economics: An introductory analysis* (1st ed.). McGraw-Hill.
Samuelson, P. A. (1988). The Keynes-Hansen-Samuelson multiplier-accelerator model of secular stagnation. *Japan and the World Economy, 1*, 3–19. (North-Holland).

Samuelson, P. A. (1989). Gibbs in economics. In D. Caldi, & G. Mostow (Eds.), *Proceedings of the Gibbs Symposium, Yale University, May* 15–17, 1989 (pp. 255–267). American Mathematical Society. (Also in J. Murray ed. *The Collected Scientific Papers of Paul A. Samuelson*, Vol.7, Cambridge MA: MIT Press, 2011).

Sandomo, A. (2011). *Economics evolving: A history of economic thought*. Princeton University Press.

Sargent, T., & Sims, C. (1977). Business cycle modeling without pretending to have too much a priori economic theory. *New Methods Business Cycle Research, 1*, 145–168.

Sato, K. (1974). *Production functions and aggregation*. North-Holland.

Sato, K.-I. (1999). *Lévy processes and infinitely divisible distributions*. Cambridge University Press.

Sato, R., & Beckmann, M. J. (1968). Neutral inventions and production functions. *The Review of Economic Studies, 35*(1), 57–66.

Sato, R., & Morita, T. (2009). Quantity or quality: The impact of labour saving innovation on us and Japanese growth rates, 1960–2004. *The Japanese Economic Review, 60*(4), 407–434.

Scheinkman, J. (2014). *Speculation, trading, and bubbles*. Columbia University Press.

Schmookler, J. (1966). *Invention and economic growth*. Harvard University Press.

Schularick, M., & Taylor, A. (2012). Credit booms gone bust: Monetary policy, leverage cycles, and financial crises, 1870–2008. *American Economic Review, 102*(2), 1029–1061.

Schumpeter, J. (1934). *Theory of economic development*. Harvard University Press.

Schumpeter, J. (1936). Review of Keynes's general theory. *Journal of the American Statistical Association, 31*(196). (December).

Schumpeter, J. (1939). *Business cycles*. McGraw-Hill.

Schumpeter, J. (1946). John Maynard Keynes. *American Economic Review* (September). (also in id. *Ten Great Economists: From Marx to Keynes*, New York: Oxford University Press).

Schweitzer, F. (2003). *Brownian agents and active particles: Collective dynamics in the natural and social sciences*. Springer.

Sethna, J. (2006). *Statistical mechanics: Entropy, order parameters, and complexity*. Oxford University Press.

Shea, J. (1993). Do supply curves slope up? *Quarterly Journal of Economics, 108*(1), 1–32.

Shiller, R. (1981). Do stock prices move too much to be justified by subsequent changes in dividends? *American Economic Review, 71*(3), 421–436.

Shiller, R. (2003). From efficient markets theory to behavioral finance. *Journal of Economic Perspectives, 17*(1), 83–104.

Shiller, R. (2017). Narrative economics. *American Economic Review, 107*(4), 967–1004.

Shimer, R. (2005). The cyclical behavior of equilibrium unemployment and vacancies. *American Economic Review, 95*(1), 25–59.

Shimer, R. (2003). From efficient markets theory to behavioral finance. *Journal of Economic Perspectives, 17*(1), 83–104.

Shimer, R. (2010). *Labor markets and business cycles*. Princeton University Press.

Shinkai, Y. (1980). Spillovers in wage determination: Japanese evidence. *Review of Economics and Statistics, 62*, 288–292.

Sinha, S., Chatterjee, A., Chakraborti, A., & Chakrabarti, B. K. (2011). *Econophysics*. Wiley-VCH.

Skidelsky, R. (2020). *What's wrong with economics?* Yale University Press.

Slutzky, E. (1937). The summation of random causes as the sources of cyclic processes. *Econometrica, 5*, 105–146.

Smith, A. (1776). *An inquiry into the nature and causes of the wealth of nations*. W. Stahan, T. Cadell.

Smith, E., & Foley, D. (2008). Classical thermodynamics and economic general equilibrium theory. *Journal of Economic Dynamics and Contral, 32*, 7–65.

Solow, R. M. (1956). A contribution to the theory of economic growth. *Quarterly Journal of Economics*, *70*(1), 65–94. (February).

Solow, R. M. (1957). Technical change and the aggregate production function. *Review of Economics and Statistics, 39*(3), 312–320.

Solow, R. M. (1979). Another possible source of wage stickiness. *Journal of Macroeconomics, 1*(1), 79–82.
Solow, R. M. (1980). On theories of unemployment. *American Economic Review, 70*(1), 1–11.
Solow, R. M. (1986a). What is a nice girl like you doing in a place like this?: Macroeconomics after fifty years. *Eastern Economic Journal, 12*(3).
Solow, R. M. (1986b). Unemployment: Getting the questions right. *Economica, 53*(Supplement), 23–34.
Solow, R. M. (1990). *The labor market as a social institution*. Basil Blackwell.
Solow, R. M. (1994). Perspectives on growth theory. *Journal of Economic Perspectives, 8*, 45–54.
Solow, R. M. (2000), *Growth theory, an exposition* (2nd ed.). Oxford University Press.
Solow, R. M. (2004). Introduction: The Tobin approach to monetary economics. *Journal of Money, Credit and Banking, 36*(4), 659–663.
Solow, R. M. (2014). Thomas Piketty is right: Everything you need to know about capital in the twenty-first century. *The New Republic*, April 22.
Sornette, D. (2003). *Why stock markets crash*. Princeton University Press.
Sornette, D. (2004). *Critical phenomena in natural sciences, chaos, fractals, self-organization and disorder: Concepts and tools* (2nd ed.). Springer.
Sornette, D. (2014). Physics and financial economics (1776–2014): Puzzles, ising and agent-based models. *Reports on Progress in Physics, 77*, 062001(28pp).
Stanley, H. E., Gopikrishnan, P., & Plerou, V. (2006). Statistical physics and economic fluctuations. In M. Gallegati et al. (Eds.), *The complex dynamics of economic interaction*. Springer.
Stansbury, A., Summers, L. (2020). The declining worker power hypothesis: An explanation for the recent evolution of the American economy. In *Brookings Papers on Economic Activity* (pp. 1–96). Spring.
Stiglitz, J. (2010). Needed: A new economic paradigm. *Financial Times*. (August 20).
Stock, J., & Watson, M. (1998). Diffusion indices. *NBER Working Paper* 6702.
Stock, J., & Watson, M. (2002). Forecasting using principal components from a large number of predictors. *Journal of American Statistical Association, 97*, 1167–1179.
Summers, L. (1986). Some skeptical observations on real business cycle theory. *Quarterly Review* (Federal Reserve Bank of Minneapolis), *10*(4), 23–27
Summers, L. (2014). U. S. economic prospects: Secular stagnation, hysteresis, and the zero lower bound. *Business Economics, 49*(2), 65–73.
Summers, L. (2015). Demand side secular stagnation. *American Economic Review, 105*(5), 60–65.
Sutton, J. (1997). Gibrat's legacy. *Journal of Economic Literature, 35*(1), 40–59.
Svenson, L. (2001). The zero bound in an open economy: A foolproof way of escaping from a liquidity trap. *Monetary and Economic Studies, 19* (S-1), 277–312.
Sweezy, P. M. (1939). Demand under condition of oligopoly. *Journal of Political Economy, 47*(4), 568–573.
Syverson, C. (2011). What determines productivity? *Journal of Economic Literature, 49*(2), 326–365.
Takayasu M., Mizuno, T., Ohnishi, T., & Takayasu, H. (2005). Proceedings of practical fruits of econophysics. In H. Takayasu (Ed.), *Temporal characteristics of moving average of foreign exchange markets* (pp. 29–32). Springer.
Taylor, J. B. (1989). Differences in economic fluctuations in Japan and the U. S.: The roll of nominal rigidities. *Journal of Japanese and International Economies, 3*(2), 127–144.
Temin, P. (2017). *The vanishing middle class: Prejudice and power in a dual economy*. MIT Press.
Thaler, R. H. (2015). *Misbehaving: The making of behavioral economics*. W. W. Norton.
Thomson, R. (1987). Learning by selling and invention: The case of sewing machine. *Journal of Economic History, 47*(2), 433–445.
Tobin, J. (1969). A general equilibrium approach to monetary theory. *Journal of Money, Credit and Banking, 1*(1), 15–29.
Tobin, J. (1972). Inflation and unemployment. *American Economic Review, 62*(1), 1–18.

Tobin, J. (1975). Keynesian models of recession and depression. *American Economic Review, 65*(2), 195–202.
Tobin, J. (1993). Price flexibility and output stability: An old Keynesian view. *Journal of Economic Perspectives, 7*(1), 45–65.
Tobin, J. (2003). *World finance and economic stability*. Edward Elgar.
Toda, A. (2009). Existence of a statistical equilibrium for an economy with endogenous offer sets. *Economic Theory, 45*(3), 379–415.
von Tugan-Baranowsky, M. (1901). *Studien zur Theorie and Geschichte der Handelskrisen in England*. Jena.
Ueda, K., & Yoshikawa, H. (1986). Financial volatility and q theory of investment. *Economica, 53*(299), 11–27.
Ushijima, S. (2004). Determination of wages and prices, and deflation in Japan (in Japanese). *ESRI Discussion Paper,* 90.
Uzawa, H. (1961). Neutral inventions and the stability of growth equilibrium. *The Review of Economic Studies, 28*(2), 117–124.
Uzawa, H. (1969). Time preference and the Penrose Effect in a two-class model of economic growth. *Journal of Political Economy, 77*(4, Part 2), 628¬652.
Van Kampen, N. G. (1992). *Stochastic processes in physics and chemistry* (revised and enlarged edition). Elsevier.
Walras, L. (1926). *Élements d'économiepolitique pure ou Théorie de la richesse sociale*. Lausanne.
Watanabe, K., Takayasu, H., & Takayasu, M. (2009). Random walker in temporally deforming higher-order potential forces observed in a financial crisis. *Physical Review E, 80*(5), 056110.
Weidlich, W. (2000). *Sociodynamics: A systematic approach to mathematical modelling in the social sciences*.
Weitzman, M. (1982). Increasing returns and the foundation of unemployment theory. *Economic Journal, 92,* 787–804. (December).
Yamada, K., Takayasu, H., Ito, T., & Takayasu, M. (2009). Solvable stochastic dealer models for financial markets. *Physical Review E, 79*(5), 051120.
Yoshikawa, H. (1980). On the "q" theory of investment. *American Economic Review, 70*(4), 739–743.
Yoshikawa, H. (1981). Alternative monetary policies and stability in a stochastic Keynesian Model. *International Economic Review, 22*(3).
Yoshikawa, H. (1990). On the equilibrium Yen-Dollar rate. *American Economic Review, 80*(3), 576–583.
Yoshikawa, H. (1993). Monetary policy and the real economy in Japan. In K. Singleton (Ed.), *Japanese monetary policy* (pp. 121–159). University of Chicago Press.
Yoshikawa, H. (1995). *Macroeconomics and the Japanese economy*. Oxford University Press.
Yoshikawa, H. (2003). The role of demand in macroeconomics. *Japanese Economic Review, 54*(1), 1–27.
Yoshikawa, H. (2010). The general theory: Toward the concept of stochastic macro-equilibrium. In B. W. Bateman, T. Hirai, & M. C. Ivlarcuzzo (Eds.), *The return to Keynes*. Harvard University Press.
Yoshikawa, H. (2013). *Deflation* (in Japanese). Nihonkeizaishinbun.
Yoshikawa, H. (2015). Stochastic macro-equilibrium: A micro-foundation for the Keynesian economics. *Journal of Economic Interaction and Coordination, 10,* (1), 31–56. (April).
Yoshikawa, H. (2020). *Population and Japanese economy: longevity, innovation, and economic growth*. JPIC.
Yoshikawa, H. (2021). *Ashes to Awesome: Japan's 6000 day economic miracle*. JPIC.
Yoshikawa, H., & Ohtake, F. (1987). Post war business cycles in Japan: A quest for the right explanation. *Journal of the Japanese and International Economies, 1*(4), 373–407.
Young, A. A. (1928). Increasing returns and economic progress. *Economic Journal, 38*(152), 527–542.

Young, A. (1995). The tyranny of numbers: Confronting the statistical realities of the East Asian growth experience. *Quarterly Journal of Economics, 110*, 641–680.

Yukalov, V. I., & Sornette, D. (2012). Statistical outliers and dragon-kings as Bose-condensed droplets. *The European Physical Journal Special Topics, 205*, 53–64.

Yule, G. U. (1926). Why do we sometimes get nonsense-correlations between time series? *Journal of the Royal Statistical Society, 89*(1), 1–63.

Zipf, G. K. (1949). *Human behavior and the principle of least effort*. Addison-Wesley.

Index

H. Yoshikawa, *Reconstruction of Macroeconomics: Methods of Statistical Physics, and Keynes' Principle of Effective Demand*, Advances in Japanese Business and Economics 3, https://doi.org/10.1007/978-981-19-5264-7

The manufacturer's authorised representative in the EU is Springer Nature Customer Service Centre GmbH, Europaplatz 3, 69115 Heidelberg, Germany. If you have any concerns regarding our products, please contact ProductSafety@springernature.com

Printed and bound by CPI Group (UK) Ltd, Croydon, CR0 4YY

30/09/2026

02225242-0003